ADVANCE PRAISE FOR

an enactment of science

"Robert W. Blake, Jr., thoughtfully portrays and interprets the considered effort and practical enactment of a dedicated science teacher. In so doing he sheds light on the practice of teaching, the multitude of influences on teachers, and the need to be ever-mindful of the central curriculum question: What is worth knowing and experiencing?"

William H. Schubert, Professor of Education, University of Illinois at Chicago; Vice-President of the American Education Research Association; President-Elect, Society of Professors of Education

"Too often, the worlds of teachers come to us in segmented pieces in scholarly journals, or through a few quotes in a book. This warm and clearly written book is an exception. Through the words of Robert W. Blake, Jr., we are introduced to Donna, an exceptional teacher, as she goes about the daily work of introducing her six-graders to the wonders of science. Too often, examples of powerful pedagogy are confined to suburban schools with middle-class populations, well endowed with resources, and rich in the kind of cultural capital schools value. Not so here, Donna works in a diverse urban environment with children others would label as 'disadvantaged.' Here we find a teacher who believes that children should have no opportunity to fail; a teacher who believes that while content has value, cultivating inquiring habits of mind is more enduring; a teacher who sees herself as enabling her students to figure out who they are, where they are going in life, and how best they can maximize their unique gifts. Furthermore, Donna values social justice as an aim of teaching with action as its outcome.

The detailed portrayal in this book unveils the complex choreography involved in balancing a given curriculum, mandated assessment, and all of the foregoing ambitions in a persuasive enactment of science instruction. For anybody who believes that teaching can be reduced to simplistic prescriptions, this book is a must read, and it is likewise for teachers who are struggling to balance their design to be humane and just in their practice with increasing demands from external curriculum and accountability pressures."

Michael O'Loughlin, Education Studies and Derner Institute of Advanced Psychological Studies, Adelphi University, New York

an enactment of science

Studies in the
Postmodern Theory of Education

Joe L. Kincheloe and Shirley R. Steinberg
General Editors

Vol. 161

PETER LANG
New York • Washington, D.C./Baltimore • Bern
Frankfurt am Main • Berlin • Brussels • Vienna • Oxford

Robert W. Blake, Jr.

an enactment of science
a dynamic balance among curriculum, context, and teacher beliefs

with a Foreword by
John R. Staver

PETER LANG
New York • Washington, D.C./Baltimore • Bern
Frankfurt am Main • Berlin • Brussels • Vienna • Oxford

Library of Congress Cataloging-in-Publication Data
Blake, Jr., Robert W.
An enactment of science: a dynamic balance among curriculum,
context, and teacher beliefs / Robert W. Blake, Jr.
p. cm. — (Counterpoints; vol. 161)
Includes bibliographical references.
1. Science—Study and teaching. 2. Acting out (Psychology).
I. Title. II. Counterpoints (New York, N.Y.); vol. 161.
Q181 .B513   507'.1—dc21   2001029120
ISBN 0-8204-5124-X
ISSN 1058-1634

Die Deutsche Bibliothek-CIP-Einheitsaufnahme
Blake, Jr., Robert W.:
An enactment of science: a dynamic balance among curriculum,
context, and teacher beliefs / Robert W. Blake, Jr.
–New York; Washington, D.C./Baltimore; Bern;
Frankfurt am Main; Berlin; Brussels; Vienna; Oxford: Lang.
(Counterpoints; Vol. 161)
ISBN 0-8204-5124-X

Cover design by Lisa Dillon
Text design and layout by Art Comp & Design Company

© 2002 Peter Lang Publishing, Inc., New York

All rights reserved.
Reprint or reproduction, even partially, in all forms such as microfilm,
xerography, microfiche, microcard, and offset strictly prohibited.

DEDICATION

To Donna and all the teachers like her, whose dedication and commitment serve to ensure the very best for all children.

CONTENTS

List of Figures ix

List of Tables xi

Foreword by John R. Staver xiii

Acknowledgments xv

Introduction 1

Chapter One
The Dynamic Balance of Teaching 7

Chapter Two
Constructing a Model of Enactment: A Balance in the Middle 19

Chapter Three
The Role of Teacher Beliefs in Enactment 35

Chapter Four
Getting "at" Enactment: Constructing a Method of Inquiry 43

Chapter Five
The Science Curriculum 53

Chapter Six
The Context 71

Chapter Seven
Teacher Beliefs: A Conceptualization of Self 79

Chapter Eight
Enactment 107

Chapter Nine
Mediation Between Self and Enactment: The Salient
Influences on Curriculum Implementation 161

Chapter Ten
An Impact on Teaching, or, an Answer to the
Question, "So What?" 183

References 195

Appendix A
Specific Data Collection and Analysis Methods 205

Appendix B
Donna's Responses to the Science Teacher
Efficacy Belief Instrument (STEBI) 209

FIGURES

Figure 1.	Criteria for moving into teams	.8
Figure 2.	Components of enactment model with interactive arrows indicating direction of influences	.22
Figure 3.	Relationship among three frameworks of enactment	.23
Figure 4.	Responses to the writing prompt "Science is . . . " made on the first day of class by preservice elementary education students	.28
Figure 5.	Lab response questions for the lab write-up	.63
Figure 6.	Student "team skills" from BSCS	.74
Figure 7.	Poster with guidelines for classroom behavior	.74
Figure 8.	Daily agenda listed on the board for Wednesday, 9/22/93	.75
Figure 9.	Afternoon list for Wednesday, 9/22/93	.75
Figure 10.	Student-generated list of ways to observe	.117
Figure 11.	Class list of ideas for "Looks Like/Sounds Like" of the new social skill	.120
Figure 12.	Social skill tally sheet used by the coach	.125
Figure 13.	Student samples of a written procedure to lab response question #2	.145
Figure 14.	Overhead list of variable types	.150
Figure 15.	Written format for science fair	.154
Figure 16.	List of material to be included in the science fair folder	.156
Figure 17.	Revised model of enactment	.164

TABLES

Table 1.	Number of Observed Classroom Lessons and Associated Times Language	.46
Table 2.	Teacher Assessment Opportunities During Labs	.64
Table 3.	Class Data Collected During "Oh Deer"	.132
Table 4.	Data Used for Practicing Graphing	.133
Table 5.	Sample Table for Keeping Track of Each Diagonal Drawn	.145
Table 6.	Sample Data Table Showing Number of Diagonals per Hexagon	.146

FOREWORD

Teaching science today can be portrayed as a tug of war. In today's school culture, the teacher stands in the middle, arms outstretched and simultaneously strapped to and holding two separate ropes. On the other end of the first rope are vocal parents, fellow teachers, school administrators, state education officials, legislators, governors, professors, business people, and others. These folks view science as a body of information. They want teachers to follow the prescribed curriculum, to present scientific knowledge through direct instruction, to test students periodically over scientific information presented, to accept the full responsibility and authority for learning, to treat all students alike, and to deal with students as a group. These folks want students to learn the scientific information presented and to compete with other students. This group represents tradition, sometimes known as the status quo. On the other end of the second rope are other vocal parents, fellow teachers, school administrators, state education officials, legislators, governors, professors, business people, and others. These people want the teacher to adapt the curriculum, to emphasize higher level comprehension and use of scientific knowledge through inquiry-based instruction, to guide students through active extended scientific inquiry, to continuously assess students' understanding, to share the responsibility for learning with students, to respond to students as unique individuals, and to facilitate a learning community in the science classroom. This group represents reform.

Regardless of which, if either, group they identify with, teachers understand this metaphor because it is a very real part of their professional lives. Teachers feel pulled and pushed by a myriad of forces in today's culture of K-12 schools, often with little or no capacity to deal with such forces. So, they shut their classroom doors and carry out their labor of love, teaching students.

The labor of love described herein is Donna's. Donna is a sixth grade teacher in an urban school setting. She is implementing an elementary science curriculum that embodies the vision of reform in elementary level science education. Robert Blake, the researcher and author, portrays and interprets Donna's labor of love in his report of an ethnographic study of her implementation of the science curriculum. Parents, fellow teachers, school administrators, state education officials, legislators, governors, professors, business people, and other stakeholders on either rope should find Blake's report valuable because it represents an all too rare bridge over the gap between research on curriculum development and research on students' learning. Blake delineates Donna's struggle to construct her own role as a teacher and her students' roles as learners as they work together in teaching and learning science, or, as Blake says, enacting the science curriculum. Donna's beliefs, background, and skills as a teacher play an important part in this process. The school science program, the students, and the school itself represent three additional parts to Donna's enactment of the science curriculum. All of Schwab's commonplaces (teacher, subject matter, learner, and milieu) are represented.

The final point that I wish to make in this foreword is in the form of a question: To what extent does Blake's portrayal of Donna's enactment of the science curriculum apply beyond Donna's own classroom? This question raises the issue of generalizability or external validity. An ethnography, as a specific kind of qualitative inquiry, has a distinct commitment toward generalizability compared to quantitative inquiry. Readers who look for the myriad of a priori conditions that form a foundation for generalization in quantitative research will not find them in the pages that follow. Readers will find, however, a full, thick, and rich description of Donna's enactment of the science curriculum. In the end, it is each reader's responsibility to examine Blake's description, to judge how closely his or her own situation matches Donna's situation, and to decide ultimately what of Donna's enactment is applicable. I think that readers will find a great deal in Blake's report that rings loudly and clearly for them. It certainly does for me.

John R. Staver
Center for Science Education
Kansas State University
Manhattan, KS 66506

ACKNOWLEDGMENTS

Putting together a book is a multi-person task, and I know there are more people that I am indebted to than I will remember. First, I would like to thank my dissertation committee, Corinna "Bunty" Ethington, Mark Smylie, William Schubert, John Staver, and Maria Varelas for their time and support in helping me in this endeavor. Special thanks to Bunty for her consistent prodding and occasional "swift kick in the butt" to move me forward.

To my fellow graduate students Beth Dohrn, Mat Jones, Maureen Daw, Becky Greenberg, Dimitra Hartas, Andy Isaacs, and Michelle Ward-Pierczynski, with all of whom I have lived and shared a portion of my life I say, in very simple terms, "thank you for being there." In addition, while they may not have lived the experience they certainly lived with it and provided unending support and encouragement. Thank you to Doug and Kathleen Nestler, JR. Dethorn, Dan and Madeline Cusick, and Frank Benchik. A special thanks goes to Dawn Szabo who worked diligently in transcribing tapes, organizing my table of contents, and always providing a reality check in her editorial comments. Dawn, I couldn't have done it without you.

To my family I thank my nephew Robbie, who taught me all over again what it means to be a little kid. To Pillow and her affection addiction as she always made me feel welcome. To my mother whose one question of "When the heck are you going to be finished?" always brought me back to reality. To my father for his unwavering support, advice, sailing excursions, and editorial expertise. To my brother Dave without whom I may not have kept current in the state of affairs in the sports world. And to my sister, Brett, who by preceding me, made life easier as she blazed a path identifying all the peaks and valleys along the way. As we shared our lives together for three years she taught me the meaning of commitment, perseverance, tenacity, and a steadfast belief that what we were doing would make a difference. Thank you. I love you all.

Finally, to Donna whose dedication to her profession, her commitment to the children, and her belief in herself makes me envious for the teacher I have always hoped to be. I thank you for an experience I shall never forget.

INTRODUCTION

To improve student learning, it will be necessary to focus on a change in teaching. . . . . How we teach is more vital that what we teach! (Yager & Lutz, 1994, p. 344)

Purpose

The purpose of this book is to look extensively at how one teacher (one considered exemplary by her peers and students) utilizes innovative science curricula and teaching practices in the specific context of an urban sixth grade classroom. Through the detailed observation and analyses of this teacher's practice we arrive at a better understanding of how she and her students manage, through a dynamic balance, the enactment of science.

Personal History and Influences upon Study

At many points during the writing of this book I reflected on why I chose to do this and I soon realized that the roots of my choice were initially planted over ten years ago. Not wanting to go into my life history (the "I was born in a one-room log cabin" story) I believe it would be useful to give readers insight into who I am and the experiences that have influenced my choice of this study. Just as it will be important to give a brief historical background on the teacher presented here, my experiences have greatly influenced the direction I have chosen to take in my career, which includes my interests in research on teaching.

Almost twenty years ago I graduated from the State University of New York at Albany with a Bachelor of Science degree in Biology. While the program had a large pre-med component and another portion of the students heading towards laboratory work, my interests were elsewhere. I was interested in animals, and particularly how and why animals did certain things (i.e., Animal Behavior). While I found my class in animal behavior difficult (writing hypothesis essays is still not my forté), I was intrigued and influenced by the works of a variety of ethologists, namely Konrad Lorenz, Niko Tinbergen, Jane Goodall, Diane Fossey, Roger Payne, and most recently George Schaller. It was particularly the work of Goodall (chimpanzees), Fossey (gorillas), and Payne (southern right whales) that captivated me. These people spent their lives trying to understand how these large social organisms interacted and what importance this had on their existence. All of their work was done in the field, within the natural setting of the organisms.

Ten years later, as I began to think about what I wanted to study concerning the teaching and learning of science, I began to reflect on what had influenced me and what was important to me. It made sense that in order to study the "how" of a teacher's practice I would have to do so in the context of that teacher's classroom. As highly complex social organisms (humans) the "natural" environment (the classroom) would be where I could get a "real" understanding of a teacher's role in the implementation of science.

The next set of influences came right after my undergraduate work. Not sure exactly how I could apply my degree to some sort of career, I decided to pursue a personal goal, one in a different direction. I became a ski teacher at Killington Ski Area, Killington, VT, and at the time I did not realize how large of an influence this experience would have on my current teaching career. As the largest ski area in the East, with one of the largest ski schools in the country, the people at Killington took their teaching seriously, and prided themselves for being innovators in the ski teaching industry. Four years later I was at Brown University earning my Masters of Arts in Teaching (M.A.T.) Biology degree.

I talk about Killington and Brown together because there were uncanny similarities between the philosophy of teaching at Brown, and the one for ski teaching at Killington. "Student as worker, teacher as coach" is what we heard on College Hill at Brown, and in ski teaching this is exactly what we did. At Brown the additional influence of Ted Sizer and the Essential Schools notion helped to continue shaping my personal philosophy of teaching. However, during my masters program one issue that endured was the question of the teacher's specific role in a "progressive" classroom. Is it okay to lecture? If so, when? When is it okay to be more direct in teaching as opposed to a strict student-centered approach? Is it okay to tell kids information or do we simply let them "construct" their own knowledge? These questions became even more important when I graduated and moved on to teaching high school science.

It was towards the end of my first year of high school teaching that I was introduced to Piagetian constructivism, and similar to the philosophy at Brown, it had the child at the center of learning. While certainly not all bad, it was when we talked about ways to help students learn science the question of the teacher's role in this process was again unclearly defined. We were to structure the learning environment and *facilitate* the students' construction of knowledge. I emphasize facilitate because this single word exemplifies jargon in science education. Generally, when used it is intended to indicate that the learning environment is "good" because it is less teacher-centered, with more emphasis on the students and their active involvement. It also seems to imply that some sort of teaching innovation is going on, and by using facilitate there seems to be little need for a description or even explanation of what is actually happening in the classroom. Facilitate told me nothing about what the teacher does in this more student-centered process of learning. I kept asking: "What role did I have as a teacher?" "Could I tell the students information?" "Could I lecture, and if so, when was it appropriate?" "When do I step back and when do I wade in to provide guidance and clarity?" "What was my role within this process of facilitating?"

Upon entering my doctoral program, I made it a point to make every reasonable attempt to partially resolve, at least for myself, the issue of the teacher's role in teaching. I wasn't satisfied with a strict traditional approach, and the opposite framework, the progressive, didn't address the fundamental issue of what the teacher does in the classroom. In fact, the literature reviews in this book are a chronology of the process I went through as I constructed my own understanding of teaching and learning, and the results, a combination of traditional and progressive framework, represent where I am today as a teacher.

The preceding discussion is important in that it brings out my personal influences on two major components of this study. The first component is instructional and relates directly to how and what the teacher does and believes in as she implements innovative science curricula. The next component was a methodological one. Humans, like other large mammals, are highly complex social organisms. Classrooms are an equally complex social environment where the interactions between teacher and students, and among the students themselves are many. Studying how a teacher sets up this environment and attempts to implement various practices will be best understood within the context of the natural classroom setting. I realize that mentioning non-human organisms in the same breath as humans may raise eyebrows among some readers. I do so only to raise an issue of method. Why is that when we want to get some understanding of other complex social organisms we do it in the "field" and yet at times, when we want to understand what teaching is all about we do not necessarily do it at the source, the classroom? Teaching is dynamic, complex, and at times difficult to understand. Scientific studies that have had the most profound impact upon my thinking have been by ethologists, those people intent on studying behavior at the source. For me, as a classroom teacher, this made sense. Thus, when I decided to study the teaching of science, I too wanted to go into the "field" and investigate what was going on. My intent was to observe a quality teacher attempting to implement innovative science curricula. In doing so I would hopefully gain some understanding of how the teacher did this. However, in order to do this I needed to pull from others, the anthropologists and the educational ethnographers who have come before me and who have built the framework for study of humans and human behavior in a natural setting.

There you have the beginning of my journey. I continue, however, to question what it means to teach and learn, especially in science, and I continue to ask what William Schubert often asks his graduate students: "What is worthwhile to know and experience?"

Organization of the Study

This book can be organized into three main sections: the introduction (Chapter one), the theoretical frameworks and method (Chapters two, three, and four); the findings (Chapters five, six, seven and eight); and the analysis and conclusions (Chapters nine and ten). Chapter two introduces the model of enactment used in this study. The four components of the model: the curriculum, the context, the teacher's beliefs, and enactment are each discussed as they relate to what the teacher does in the classroom. The three conceptual frameworks of teaching (traditional, progressive, and middle ground) are discussed under the enactment heading. Chapter three discusses teacher beliefs and the role they have in classroom practice. Chapters four, as well as Appendix A, discuss the methodology and the theoretical framework underlying an ethnographic case study approach.

The next major section consists of the findings of the study, with one chapter for each of the four components of the model. Presented in each chapter will be those factors that may influence the teacher's practice. Chapter five contains detailed information about the specific science curricula used by the teacher in this study. Chapter six provides detail relating to the classroom and the school context. Chapter seven, taken mainly from formal interviews, provides an in-depth description of Donna's belief structure as it relates to teaching and learning in general and specifically to teaching and learning in science. Finally, Chapter eight presents the teacher's actual classroom practice. This chapter will provide exam-

■ Introduction

ples supporting the original intent of the study (movement between traditional and progressive), as well as the emergent theme of social skill use and how it fits into the overall scheme of enactment.

The final section contains the analysis, conclusions, and implications. In Chapter nine a cross component analysis, I will discuss how enactment can be explained in relationship to the curricula, the context, and the teacher's beliefs. A concentrated focus will be on the interplay between the teacher's beliefs and the enactment of innovative science curricula. Chapter ten relates how this teacher teaches and discusses the implications of how this model of enactment, as well as her practice, may be useful for others as they reflect on their ways of teaching science.

Knowing full well that a single case study has little value in its generalizability, I hope that each reader may find parts of this story that resonate with his or her own classroom practice. I believe that teachers are at the heart of any success in educational reform, and understanding what they do is key to positive change in classroom practice. For me, "How" they teach is more important than what they teach.

CHAPTER ONE
The Dynamic Balance of Teaching

Day One

Everyone is asked to quietly get out of his/her seat and to stand on one foot. After five seconds, Donna asks, "Okay, things are going on as you're standing on one foot. What are you doing? What happens?" Hands are waving in the air, and bodies sway back and forth, poised to topple over. Donna continues to probe the students for their ideas of what is happening.

> Sometimes you shake, but what are you doing? Are you absolutely still, or are you making some movements to stay still? Are some of you moving your ankle back and forth so you maintain that balance? Okay, and even though you're standing here, if you start off balanced you might have to make some adjustments to remain balanced. Right? You make little shifts of movement. I see some of you moving around like tightrope walkers. (All quotations are taken from classroom audiotape recordings)

Her students are having difficulty standing on one foot, and there are giggles and laughs as they try to maintain their balance. A number of them put their foot down or lean on their desk to stabilize themselves. After a few seconds more, Donna tells them to be seated, and once they are seated, she goes on to explain the first theme for the year in science.

> Okay, what you've just demonstrated is a major theme we are doing ... to be talking about this year, is the theme emphasizing Dynamic Balance [she writes this on the board]. Dynamic means that things could be in motion or that they might be changing. And we're going to be investigating the way that systems make slight changes in order to remain in balance. Just like you, to remain standing had to shift side-to-side, systems also make slight changes so that they can remain balanced to nature.

She then continues to explain the second theme that they will concentrate on during the school year.

> You might also, another thing we're going to be investigating is, [pause], decision making process and how decisions, what kinds of decisions you might need to make in order to remain in dynamic balance. You had to decide somewhere, your brain had to make decisions to move your ankle around so you wouldn't land flat on your face. There were some decisions that went on there.

After a brief introduction to the themes, Donna tells them that they can now move onto the "fun stuff," but she explains that first they will need to practice a social skill. One such skill demands that they move from their "home-base" to their science teams "quickly and quietly." Home-base is the regularly assigned seat for each student. This seat is considered "their" desk, where they are able to leave their school belongings: books, journals, pens, pencils, markers, and other possessions. Home-base is where each student sits at the beginning of the day and when they come back from lunch. Certain activities are usually conducted at home-base, including reading and writing workshop and math lessons. However, many of the daily activities are done in assigned teams of three. This includes science, social studies, and sometimes math (math/science labs are done in pairs). Thus, moving from home-base to a new team in a quiet, efficient, and organized manner is crucial for effective management of the class, and thus, is an important social skill for them to learn.

In assessing the last time they went into teams, the class responds that they did an "awful" job. Donna: "All right, so that would be something we need to practice on." Displaying a list on the overhead, she goes over each step the students will need to think about as they move into teams (Figure 1).

---

1) Take what you need.
    Journal
    Book
    Writing Utensil
2) Know where you're going
3) Don't stop and talk on the way. Go directly to your station.
4) Stay quiet and turn your attention up front.

Figure 1. Criteria for moving into teams

Each station is numbered, and each team member is to know the station that she or he will report to. To ensure a smooth transition, Donna suggests that each member make eye contact with the other members to make sure they know who is on their team and where they will report. Finally, and because a few stations may not have chairs, she suggests that they check beforehand to see if they need to bring one. To hold them accountable for the move, Donna keeps time of each transition, and expects that they can do it within thirty seconds. Donna: "I'll be looking for evidence of these behaviors as you get into your groups quickly and quietly . . . and I'm going to be evaluating you" on moving into teams. After the move, which they made in thirty-five seconds, Donna provides her input.

Much improved. Much better. Okay. Still need to work on a couple of people stopping and talking. I did hear some conversation. I also heard someone calling for the group number. That isn't going to help. They'll get there. Try to find another way, maybe make eye contact, but that was much improved. If we practice it more and more, we'll get better and better.

Originally a teacher of high school Russian in the early 1980s, at the time of this study Donna was a sixth grade, self-contained teacher in a large urban school system. At Oppenheimer[1] Magnet School, which has a maintenance bilingual (Spanish) program, Donna was responsible for all subjects except foreign language. Her move from high school to elementary school necessitated that she take a different approach to teaching, one that would more directly involve her students. Practicing the use of social skills, which include the organized movement to and from working teams, as well as learning methods of positive social interaction within teams, was emphasized in the conduct of her entire classroom. Within the context of science as a way of investigating and problem solving, social skills were seen as the integral component as students learn to work in a collaborative manner. These goals were not only built into the science curriculum (Biological Science Curriculum Studies, 1992a, 1992b), but they were also included in all other aspects of the science activities designed by Donna. In fact, the main reason that she followed the science curriculum, *Science for Life and Living* (BSCS, 1992b), so closely was because the curriculum paralleled her beliefs of what it means to teach and organize the classroom.

Donna was chosen (or should I say she chose me?) for this study for two main reasons: She was implementing innovative science curricula over a long period of time, and she assured me she would be at ease with a stranger observing her. She indicated that she did not care whether or not I was being evaluative. She would do what she felt was best for her kids. Ultimately, with Donna, "what you see is what you get."

Original Intent

The purpose of this study was to investigate the enactment process of one exemplary teacher as she implemented an "innovative" (BSCS, 1992b, p. T-7) science curriculum in her sixth-grade classroom. Initially, the purpose was to see how this teacher mediated between "traditional"

---

[1]The names of the school, principal, and children are pseudonyms, with the children choosing their own pseudonyms.

and "progressive" frameworks of teaching science: traditional being more teacher-centered instruction and progressive being child-centered (The details of each framework will be examined more closely in Chapter two). Specifically, the investigation was to document instances of when, and why, components of a more "traditional" or a more "progressive" kind of teaching were used within the enactment of science instruction. The idea was to gain a better understanding of a "middle ground" between traditional and progressive types of teaching and to gain insight into why the teacher structured the learning this way.

Implicit with understanding the why of the teacher's actions came a need to delve into the teacher's general beliefs of what it meant to teach and learn generally and then specifically what it meant to teach and learn science. Linking these beliefs to a detailed description of enactment might allow one to come to a clearer understanding of the role beliefs play in practice. Also of interest were the potential influences that the curriculum and the classroom context had on teacher beliefs and enactment. The view was taken that any of these influences would be reciprocal in nature, or bi-directional in the effect.

The rationale for this study was based on the notion that very little is actually understood about classroom practice and the teaching of science in a self-contained setting. Tobin, Tippins, and Gallard (1994) suggest that "investigations of exemplary teachers might provide models for other teachers to emulate in order to enhance science learning of students" (p. 52). While this study may provide such a model, it is not intended to be generalized across a wide variety of contexts. The issue of context, therefore, is an imperative component as one judges the study's importance and usefulness. Tobin and Tippins (1993) explain that the purpose of interpretive research

> ... is to inform others of what has been learned from a study ... [and] ... not to convince readers of the generalizability of what has been learned but to provide sufficient details of the contexts in which the theory is embedded and to enable readers to decide on the extent to which what has been learned can help them meet their goals. (p. 19)

Therefore, the strength of this story may be in the rich detail, which is described in a specific context and which will allow others to make connections and to determine its value for them.

Emerging Theme

As the study progressed an emerging theme materialized. This theme focused on the *social nature of enactment*. Seeing this, I began to question how the teacher and the students set up and operated within the social organization of the classroom. I was specifically interested in how the teacher, within the process of enacting innovative science curriculum, coordinated social components of learning. What soon became clear was that she relied heavily on training her students to develop social skills that she believed were necessary for collaborative problem solving. Hence, *socializing* her students into science was an important part within the enactment process. The obvious questions thus became "How did she do it?" and "Why?" Again, understanding this theme meant linking her beliefs to her practice, and if socialization into learning science is an integral element of this teacher's repertoire, where does it fall in relation to different styles of teaching—*traditional, progressive,* or *middle ground*? If learning is truly a social process with the social nature of collaboration and negotiation of meaning imperative for learning, then understanding how students learn how to *socialize*, and how the teacher structures this socialization, may be crucial in understanding science instruction.

The substantive focus was, therefore, on the teacher and how she organized and implemented science curricula, paying close attention to the curricular decisions and choices she made as they related to her beliefs of what it meant to teach and learn science. The need to socialize students as they studied science and how this was played out in the classroom through her daily actions and interactions with the students, by means of direct instruction, modeling, and/or guided learning, became the focal point of the investigation.

The Problem

The problem and the purpose of this study are linked together under the umbrella of teacher practice, with the focus of the study centering on what one teacher does in the teaching and learning of science. In order to understand clearly the purpose of the study, one must have the salient terms defined. As I define the terms I will also describe the sources of these terms and discuss how they will be used in this study.

Identifying Frameworks of Teaching

Contemporary science education is in the midst of what Kuhn (1970) terms as a "paradigm shift" (J.R. Staver, personal communication, April 3, 1993). However, if the teaching and learning of science is undergoing a change, then it is crucial to formulate a better understanding of what happens within the classroom as a result of these changes, and specifically as the instruction relates to the teacher's role within the process of teaching science. Tobin, Kahle, and Fraser (1990) suggest that we need to gain insight into ". . .what teachers and students do, why they do what they do . . ." (p. 4) during science instruction. They suggest that as a result we will be better able to address enduring problems of science education, including an understanding of students' apathy toward science and a general decrease in their ability to understand basic scientific concepts. This problem of knowing what happens in the classroom seems more acute at the elementary and middle school levels than at the secondary level, where most classrooms are self-contained and the teacher is responsible for all subjects. If we are to improve student learning in science, we need to understand clearly what good teachers do. Also, by improving learning at the lower grades, so that students develop skills, knowledge, and habits of mind (American Association for the Advancement of Science [AAAS], 1989) at an early age, we help foster growth in the understanding and enjoyment of science, and come to understand the teaching that promoted such improved learning.

Part of the problem of understanding the idea of change in science instruction is in identifying practices that are different from "traditional" practices (traditional being practices that emphasize rote learning, teacher centered lecture and note taking, and short answer multiple choice type exams). The current literature of science education is well documented with descriptions of this practice (Cosgrove & Osborne, 1985; Goodlad, 1984, cited in Oakes, 1990; Harms & Yager, 1981; Novak, 1991; Shymansky & Kyle, 1992; Tobin, Briscoe, & Holman, 1990; Yager, 1991b; See Chapter two for a detailed description). Overcoming traditional science teaching, the norm in many classrooms, is considered a major goal of any reform in science education (AAAS, 1989, 1993; National Committee on Science Education Standards and Assessment, 1993a, 1993b, 1994a, 1994b; National Research Council (NRC), 1990; and National Science Teachers Association (NSTA), 1992; United States Department of Education, 1993).

"Progressive" methods of teaching science are considered counter to traditional teaching. Here the emphasis is placed on the child's interests with learning a direct result of induction and involvement with activities, experiments, and problems. Many terms are used interchangeably to describe this type of instruction. A few are *hands-on, process science,* or *discovery learning.* In this study, the term Progressive, most commonly associated with the work of John Dewey (1938) and his critique of the child-centered movement of the Progressive Era, will be used to represent a pedagogical framework that is in opposition to the practices of traditional teaching. It is important to note that Dewey has often been misinterpreted as strictly promoting a child-centered framework for learning. On the contrary, while Dewey did advocate teaching and learning that was opposed to traditional practices, he advocated an environment that directly involved the teacher in the learning process while also placing responsibility on the students for their own learning. This direct involvement of the teacher embedded in student-centered learning provides a pedagogical framework that lies between traditional and progressive teaching.

It is important to identify the terms "traditional" and "progressive" as pedagogical concepts that are at polar opposites. By doing so, one finds room for other frameworks that constitute the space between the two and make up some "middle ground" of teaching and learning science. The specifics of a "middle ground" are discussed more fully in Chapter two.

Exemplary Science Teaching

Tobin, Kahle, & Fraser (1990) indicate that exemplary teachers of science are those who ". . . focus on students' learning with understanding, use strategies to encourage students to engage in higher-level cognitive tasks and maintain a classroom environment conducive for learning" (pp. 3-4). Tobin and Fraser (1988) maintain that these teachers promote the development of ". . . concepts, inquiry skills and positive attitudes" toward science (p. 369). In describing the roles for those who teach science, Osborne and Freyberg (1985) provide a summary of what they've seen in their observations of exemplary science teachers.

> What has stood out, in our minds, about the successes we have seen is the way in which the teacher concerned conveyed to his or her pupils a sense of continuing involvement in exploring some small corners of the world, as though there was always the possibility of something new and interesting turning up. It was their enthusiasm for enquiry which was so contagious, and which we regard as one of the most important goals of science teaching. (p. 99)

Tobin et al., (1990) suggest that quality teachers are committed to changing their practice, have a vision of what they want the practice to be, and personalize that vision to fit both their needs and those of their students. In the book *Best Practice*, Zemelman, Daniels, and Hyde (1993) have identified ". . . thirteen interlocking principles, assumptions, or theories . . ." (p. 7) that are associated with exemplary practice, a movement away from traditional teaching. In this list they include terms like "child-centered, experiential, reflective, social, collaborative, and constructivist" to describe those teachers who exemplify *Best Practice* (See pp. 7-8 for the entire list and associated descriptions).

To round out this inventory of exemplary characteristics, Raizen and Michelson (1994), in using information from the 1991 National Board of Professional Teaching Standards, provide what amounts to a laundry list of the qualities that "effective" science teachers possess. This list includes a dedication to making knowledge accessible to students; a belief that all students can learn; treatment of all students in an equal manner; ability to recognize and account for individual differences; ability to adjust practice to fit the needs of the students; understand learning and development of the child; provide a wide range of role models for the students; provide a wide variety of experiences; be proactive instead of reactive; be skilled as a coach to help the students move forward as opposed to backward; and to promote the pursuit of lifelong learning. Finally, while Tobin and Fraser (1991) identify features of exemplary teachers, and suggest that continued research can help build theoretical models for teacher change, they do emphasize that the notion of exemplary may be in the "eye of the beholder" (p. 231).

In this study the term "exemplary" was used only to locate a teacher to observe. From comments made by her peers, by her principal, by the parents of her students, and by her students themselves, Donna is considered exemplary. In addition, by the end of the study, and from my own experience and observations, I also consider Donna an example of an exemplary teacher. However, building a case for exemplary practice is not the intent here, and, thus, it will be up to the reader to decide whether or not what Donna does in the classroom is considered excellent teaching.

Implementation and Enactment

The terms *implementation* and *enactment* often appear together. The terms, however, are not interchangeable and do have different meaning with enactment considered a component of implementation.

Implementation often refers to how the teacher puts into action a certain curriculum as she structures the learning environment for the students. In their analysis of curriculum implementation literature, Snyder, Bolin, and Zumwalt (1992) indicate that implementation seems to occur in three general approaches: *fidelity, mutual adaptation,* and *enactment.* I will briefly discuss each here.

Snyder et al. (1992) refer to *fidelity* as a linear process, where the curriculum is created outside the classroom and is passed down to the teacher who then implements it the way the developer had intended. In this sense, the teacher has no real control over what she teaches, and if she is faithful to the curriculum, she has no control over how she teaches. Such an approach, often considered "teacher-proofing" the curriculum, assumes that teachers are not capable of making decisions about their own practice. In this sense, outsiders, usually scientists, created the curriculum and passed it down to the teachers who were then expected to implement it exactly as written. Referred to as a "technical-rational" approach to curriculum, this approach was dominant in the science curriculum projects of the 1960s (see Blades, 1997, for an in-depth discussion on technical-rational, top-down approach to science curriculum design and implementation).

Mutual adaptation is a process by which the teacher uses a curriculum but adapts it to fit the needs of her[2] students. In doing so, the teacher may change and alter the material as she sees fit. Snyder et al. (1992) indicate that one interpretation has fidelity and enactment at polar opposites, with mutual adaptation somewhere in between the two. Others, however, see the lines between mutual adaptation and enactment as blurred, without a clear delineation between the two.

In curricular enactment, the teacher *and* the students play a direct role in defining the curriculum. "From the enactment perspective, curriculum is viewed as the educational experiences *jointly* created by student and teacher" (Snyder et al. 1992, p. 418, emphasis added). In this perspective, the teacher's role becomes more active in creating experiences, but the children also guide and direct these types of experiences. Snyder et al. (1992) liken this to kindling a fire, where the teacher's job is to provide the initial fuel and the bellows that will ignite a flame in the child's mind. It is the child who then seizes the opportunity to continue to fuel the fire.

---

[2]To avoid the messiness of using her/her or she/he to account for all possibilities of teachers, in this study I will consistently refer to the female gender in all references to a teacher.

Here the term *enactment* will refer not only to how the teacher implements the curriculum but also to the influences that the children have on that process. Thus, the lines between mutual adaptation and enactment do become blurred. In addition, enactment includes the social processes of how the teacher and the students construct the learning environment within the specific context of science instruction. This interpretation of enactment parallels Erickson's (1985) notion that teaching is a process of creating a culture of shared meaning between teacher and students and the building of a classroom culture constituting a learning environment. *Implementation* is merely putting the designed curriculum into action. *Enactment* is the process by which the students interact with the teacher as well as how they interact among themselves and with the curricular material.

Unlike Snyder et al. (1992) who define enactment "as the educational experiences *jointly* created by student and teacher" (p. 418, emphasis added), Erickson (1985) views construction of the enacted curriculum mainly ". . . but not exclusively, the responsibility of the teacher as the instructional leader" (p. 133). As the adult, the teacher has primary responsibility for structuring and influencing the day-to-day features of the classroom. How she goes about this, how she manages to create an environment that engages the students in a learning process, one in which they have a stake and a genuine interest of continued participation, is key in understanding the enactment process. This distinction is useful to consider as this story unfolds.

Summary

The central theme of this book is on teaching. In bridging the gap between research on curriculum development and what students do in the classroom (outcomes and achievement), I realized the necessity of coming to understand what and how teachers teach. In general, a few key questions that could be asked are what decisions does a teacher make as she structures the classroom environment for the learning of science; how does her personal philosophy of what it means to teach and learn science influence the creation of the classroom structure; what influence do the external science curricula have on teacher decisions; how do her philosophy and the curricula affect the decisions she makes and what she does in the classroom; and what roles does the teacher play in organizing learning and interacting with her students as they are engaged in a learning process?

These questions, and many like them, form the foundation for my ethnographic inquires as well as for my own reflections on teaching. I hope that you too will generate questions not only from this story but also from your personal inquiries into what influences the *how* and *what* of your practice as you engage students in learning, especially in science.

CHAPTER TWO
Constructing a Model of Enactment: A Balance in the Middle

Yager and Lutz (1994) claim that in science education "changes in teaching, not curriculum, are the key to effective reform" (p. 338) and that it is a mistake to focus continuously on "what" to teach. By focusing on the necessity of studying exemplary teachers in science, Tobin and Fraser (1988) assert "that a great deal could be learned by watching the best teachers in action and speaking with them about their teaching (p. 369). Through the careful observation

> ... of successful science and mathematics teachers, teaching practices could be described, analyzed, and presented in a form that would be useful for teachers and teacher educators and would contribute to the improvement of science and mathematics teaching (p. 369).

However, there seems an uncertainty about classroom practices and what teachers do to engage students in science. F. James Rutherford, Chief Education Officer of the American Association for the Advancement of Science (AAAS), commented that "we have no idea how teachers teach science in the classroom" (Communication made on radio news broadcast, WXRT, Chicago, IL, 11/93), and Yager (1991a) maintains that the central failure of any "fundamental reform" in science education is primarily the responsibility of teachers because ". . . they are major forces for maintaining the *status quo* in curriculum" (p. 91).

This viewpoint, the need to study teacher practice because of a lack of understanding of what actually happens in the classroom, provides the impetus for finding out what teachers do during instruction, and for identifying influences upon their practice. By doing so we would add a crucial component to our understanding of what it means to teach and learn science.

Studying how a teacher teaches, however, must include more than just a description of her practice. What a teacher believes, as it relates to teaching and learning, will also play a vital role in the structure and organization of her classroom (Clark & Peterson, 1986; Fraser, 1994; Kagan, 1992; Munby, 1982, 1984; Nespor, 1987; Tabachnick, Popkewitz, & Zeichner, 1979; Tobin, Tippins, & Gallard, 1994). Investigating teacher beliefs and linking these beliefs to action in the classroom will thus provide insight into why and when teachers do what they do. Regardless of curricular intent outlined by authors, teachers tend to rearrange the curriculum to fit the likes and needs of both themselves and of their students (Olson, 1981), thus suggesting a strong connection between teacher beliefs and the implemented curriculum (Hawthorne, 1992; Munby, 1984).

Understanding the classroom context and looking at a teacher's practice is useful for discerning the potential influences that the particular situation has upon teacher decisions. The setting in which the study is embedded will also be useful to readers as they make connections between theory and practice and create meanings that are useful for their own instruction. The ultimate strength of any study will be in how each individual reader judges the relevance and links the descriptions to her own understandings and experiences of what it means to teach science.

Conception of a Model of Enactment

Schwab's Model

Schwab argued, in *The Practical: Translation into Curriculum* (1973/1978), that four components of educational thought (the learner, the teacher, the milieu, and the content) are vital in the construction of any curriculum. He also stated that because of an overreliance on theory, and an inability to combine and use theoretical constructs to address (practical) issues of school and schooling "the field of curriculum is moribund" (p. 287, 1978). In his words "theoretic" knowledge is considered useful for "general or universal statements" where one overarching premise is considered true and "correct."

> The subject matter of the theoretic is always something taken to be universal or extensive or pervasive and is investigated as if it were constant from instance to instance and impervious to changing circumstance (p. 289).

The practical, however, "... is a *decision*, a selection and guide to possible action" (p. 288). In this view the practical is connected to an immediate need in which there is "no great durability or extensive application." The subject matter of the practical is thus considered to be concrete, and "susceptible to circumstance" (p. 289), which is dependent upon a number of influences on school and schooling (for example, the students, the teacher, the milieu or classroom and school context, and the curriculum). In adopting this view of a practical approach to curriculum, teachers tend to focus on the reality of the moment, the here and now of teaching and learning, as opposed to a generalizable, overarching theory of curriculum design. A fifth component suggested by Schwab, and one considered vital in any process of curriculum revision, was the act of *curriculum making* itself: the act of designing what to teach and the process of implementation.

Schwab's model of the practical and its relationship to the act of curriculum planning and design, is helpful here because his arguments have added important insights to this study. For example, Schwab did not specifically raise the issue of curriculum implementation or enactment, the bridging of the teacher's theoretical constructs of what it means to teach and learn to her actual classroom practice. I was intrigued, however, by the parallel nature of both his model and the one that will be presented here. By adding an implementation component and changing the emphasis from the sole task of curriculum design to one of curriculum implementation and enactment, the model in this study addresses the pragmatic aspects of classroom practice and how a teacher implements curriculum in the day-to-day setting of her classroom.

Clark and Peterson's (1986) model of a teacher's thought and action also influenced, to a lesser extent, the model of this study. Their model is separated into two parts: a Teachers' Thought Processes and a Teachers' Actions and Observable Effects. The importance of this model lies with the suggestion that the influences within each of the individual domains (for example, in the teachers' theories/beliefs and teacher planning), as well as between domains, are embedded in a reciprocal interactive process. The argument for bi-directionality is that traditional process-product research, which portrays a uni-directional, linear path of teacher thought processes, is not sufficient to account for the reality of the interactive effects upon teacher thought processes and teacher action. Thus, as teachers plan and enact curricula, they do so in relation to beliefs about their own teaching, as well as in response to actual occurrences in the classroom.

Models for This Study

The primary model used here contains four interactive pieces: the curriculum (the actual written component to be used during science instruction), the teacher's beliefs as they relate to general teaching and science teaching, the classroom context or "milieu," and the process of curriculum implementation or enactment (Figure 2). This model also incorporates an interactive nature, as represented by the double arrows as suggested by Clark and Peterson (1986). The learner, specifically omitted as a separate component in this model, becomes embedded within the context of the classroom and surely influences the teacher and her beliefs as she implements the curriculum. While no research on teaching can artificially separate the learner from the teacher, the intent here was to understand better, from the teacher's perspective, how these pieces interact and influence each other within the process of teaching and learning science. The main focus of the model is, therefore, on the enactment of science instruction as represented through the teacher and her beliefs as they play an influential role in this process.

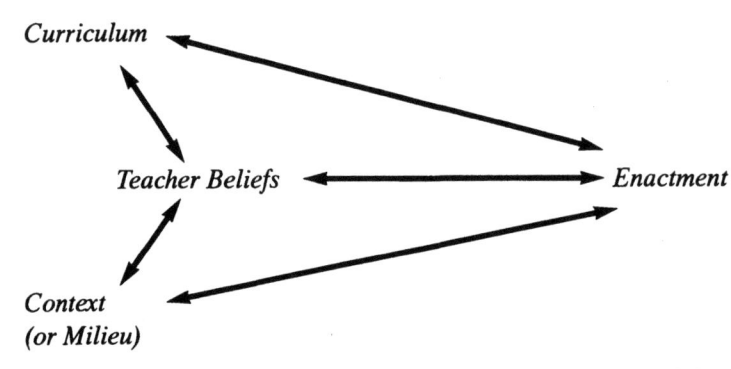

Figure 2. Components of enactment model with interactive arrows indicating direction of influences

A secondary model represents the relationship among the three frameworks of science teaching described here: traditional, progressive, and middle ground (Figure 3). The middle ground is seen as containing components of the other two frameworks, the extent of which is dependent upon the teacher and the relationship among her beliefs, the curriculum, and the context, as represented in Figure 2.

Understanding the three frameworks of teaching science will be useful as we look more closely at the actions of this teacher. The following section will include a discussion of each framework and will, therefore, represent a more traditional literature review, the purpose of which is to provide theoretical insight into these frameworks and to suggest how they can be used to construct an understanding of enactment.

Chapter Two: Constructing a Model of Enactment

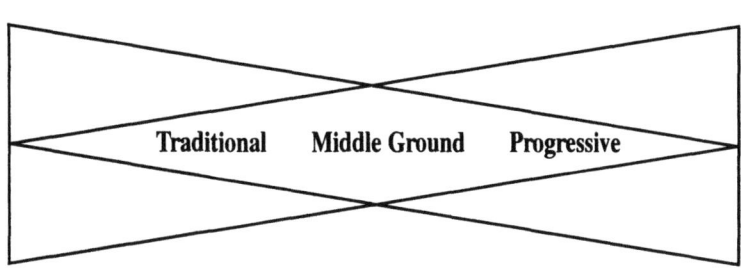

Figure 3. Relationship among three frameworks of enactment

Frameworks for the Enactment of Science

Traditional Science Education

The dominant pedagogical paradigm in science education is the behaviorist, empiricist philosophy (Goodlad, 1984, cited in Oakes, 1990; Novak, 1991; Shymansky & Kyle, 1992; Yager, 1991b). Content knowledge within this paradigm is generally considered to be finite and existing "out there" for students to access from the ideas transferred to them by the teacher: a conduit metaphor (Shymansky & Kyle, 1992; Tobin, Briscoe, & Holman, 1990). Within this paradigm "The facts of science . . . are given an aura of total objectivity" (Lemke, 1990, p. 137), and teaching relies predominantly on lecture, whole group instruction, individual competition, "cook-book" type laboratory exercises, and rote learning of definitions and facts that may be out of context and isolated from real-life situations. Little emphasis is placed on the "development of higher-order thinking skills" (Tobin, Kahle, & Fraser, 1990, pp. 1-2), and approaches to learning within this paradigm have been considered "rigid and uncreative" (Hilliard, 1989, p. 68), oriented toward abstract teaching that is devoid of any practical application. These approaches require students to match objectively their observations with what is considered to be preexisting reality.

Much of literature in the 1990s on science education has proposed a drastic change in the curricular format of science (AAAS, 1989, 1993; National Committee on Science Education Standards and Assessment, 1993a, 1993b, 1994a, 1994b; National Research Council (NRC), 1990; and National Science Teachers Association (NSTA), 1992; United States

Department of Education, September, 1993). One goal advanced in these reports is to create science instruction that allows for more student involvement and interaction as they investigate relevant topics and issues. By altering the curricular approaches and changing classroom practices, teachers enhance the opportunities for students to learn science, and the quality of such opportunities will not only enhance learning but will also be made available to all students, such as those of differing ethnic, cultural, and linguistic backgrounds. (See the following authors for more details: Johnson, Johnson, Scott, & Ramolae, 1985; Linn & Hyde, 1989; Oakes, 1990; Peterson & Fennema 1985; Skolnik, Langbort, & Day, 1982; Tannen, 1991, Banks, 1988; Goodlad & Oakes, 1988; Hilliard, 1989, 1992; Lee, 1992; Oakes, 1985; O'Loughlin, 1992; Shade, 1982; Shymansky & Kyle, 1992; Sizer, 1985, 1992).

One aspect missing from much of this literature is a clear description of what is good teaching and a description of the role of the teacher as she implements new and different frameworks for learning science. One reason for this deficiency is most likely because teaching is a localized endeavor, where a "one size fits all" technique will not be applicable to the variety of different contexts each teacher faces. Thus, collecting more information about how an exemplary teacher teaches, especially in contrast to a traditional framework, is a central component of this study.

Progressive Science Education

The word "Progressive" can have a variety of connotations as it applies to education. Defined in the *American Heritage Dictionary, New College Edition* (Morris, 1978), progressive refers to any educational reform that is considered "liberal." It is also used to describe a teacher, or a school, whose educational practices are

> . . . influenced by a theory of education characterized by emphasis on the individual needs and capacities of each child and informality of curriculum (p. 1045).

As it applies to teaching and learning that is nontraditional, progressive represents a framework embracing a more humanistic approach, where students are more directly involved and have a greater responsibility for their learning than they do in a traditional approach. For ease of comparison, *progressive* in this study will refer to practices that are at polar opposites from the traditional framework, those experiences in which student interests and needs are the center of focus and where the teacher's role is to structure the learning and act as a "guide" or "facilitator."

John Dewey and Progressivism. In *Experience and Education,* John Dewey (1938) compared traditional practices with those implicit in a "new education," one considered progressive. Here, Dewey criticized those in progressive schools for moving in a direction opposite to the traditional approach, where they proceeded "negatively or by reaction against what has been current in education" (p. 22), rather than by building a positive framework to counter traditional practices. His critique was that the progressive schools:

> ... tend to make little or nothing of organized subject-matter of study; to proceed as if any form of direction and guidance by adults were an invasion of individual freedom, and as if the idea that education should be concerned with the present and future meant that acquaintance with the past has little or no role to play in education. (p. 22)

In this sense *progressive* implies that beyond setting up the learning environment, there is very little teacher involvement within the learning process. The environment is truly student-centered where children are allowed to pursue individual interests and open-ended activities on their own. From these practices comes an interpretation often described by teachers as *discovery learning,* in which multiple outcomes can be had as students "discover" meaning for themselves. Such an approach is at extremes with the notion of "learning by transmission," which is closely associated with traditional practices.

Conventional Constructivism. Constructivism (see, among others, Lawson & Renner, 1975; Novak, 1991; O'Loughlin, 1992; Padilla, 1991; Shymansky & Kyle, 1992; Tobin & Tippins, 1993; Yager, 1991b) is a theoretical framework in contemporary science education in which knowledge and meaning-making are individually constructed. Here, learning takes into account the student's prior experiences and conceptual understanding, and meaning is created (or constructed) through a process of "negotiation and consensus building" (Tobin & Tippins, 1993, p. 4). Similar to her role in a progressive approach the teacher's role in this framework is to create the learning environment and then to structure the learning opportunities that will help the student build new meaning, relative to these current and previous experiences.

In the conventional sense of a constructivist framework, "negotiation and consensus" create a state of internal disequilibration for the individual learner. In this framework, dialogue with others, while a social endeavor, creates an internal conflict and prompts the individual to reconsider her conceptual understanding. Here, the construction, or reconstruction, of meaning is done by the individual, without "transmission" or any real negotiation or discussion of information through peer or teacher interactions.

The similarities between a progressive and a constructivist approach are apparent. If knowledge and meaning making are individual constructs, with the teacher's role limited to organizing learning experiences, then the notion that students create their own meaning is similar to learning in a progressive framework, where students work independently of teacher direction or without "imposition from above and from outside" (Dewey, 1938, p. 18).

One critique of this model of a constructivist framework is that it does not fully account for the social nature of learning. Here, discourse is intended to create internal cognitive conflict, which the individual uses as a catalyst for constructing new knowledge or new understanding. Socializing, therefore, is meant *only* for individual construction of knowledge and has little role in the collaborative nature of meaning-making. However, if learning via a constructivist framework is truly to take into consideration each individual's background and experiences, then it can be argued that the social, cultural, and historical influences of the students lives must also be considered (Driver, Asoko, Leach, Mortimer, & Scott, 1994; Edwards & Mercer, 1987; O'Loughlin, 1992). If these additional factors are taken into account, then what each student brings to the classroom is based on her subjective experiences, and thus, negotiation and consensus through a social process are imperative for the construction of understanding. If this is true, then in order to create joint understanding within the classroom, all individuals (including the teacher) must learn to discuss effectively and to negotiate meaning that helps to create *common knowledge* (Edwards & Mercer, 1987), and thus, common meaning. The teacher's role, therefore, becomes more demanding, as she becomes a social leader, one who models and practices the socialization process of meaning-making.

Creating a Middle Ground. By linking this current interpretation of constructivism with the progressive framework described by Dewey (1938) it is intended to help dichotomize progressive and traditional practices, with constructivism being more akin within the progressive framework. This division then lays the foundation for the conceptual framework of this study and provides room for another framework of teaching that is situated between these two extremes. For example, if knowledge is considered as an individual construct, based on the subjective nature of previous and current experiences, then a traditional framework, centered on the transmission and the objective nature of knowledge, would be inadequate as a framework for teaching and learning. However, if students collaborate and construct meaning in a social setting, necessitating a more active role for the teacher as she helps to clarify, guide, and mediate the discussion, then a *laissez faire* approach, associated with both a progressive and constructivist framework, would also be inadequate for student learning. What may be possible, therefore, is the combination of components of both the traditional and the progressive, used by the teacher in situations that are deemed necessary for learning by her students. Such a combination leads to what I call the *Middle Ground*, one that is different from both traditional and progressive (possibly combining elements of each), and one that may be useful in the formulation of an alternate theory of instruction for science education.

The Middle Ground: Redefining the Role of the Teacher

The Process and Product of Science as a Discipline

In discussing a middle ground of teaching science, it is initially helpful first to present two views of science and to link these views to teaching practices in the classroom. The two most common notions of science are first, as a body of knowledge, and second, as a process of inquiry. The former notion, most prevalent in traditional schooling, views science as a body of facts, terms, and a study of the specific disciplines. Such a view of science, even for adults, is often considered boring, hard, or scary. For example, when asked to respond to the writing prompt "Science is . . . ." preservice elementary undergraduate and graduate students in an elementary science methods class supported this traditional notion of science as well as an idea of what science education is in our society (Figure 4). The second notion of science, that of a process, a way to investigate and ask questions, and a way of learning how to learn, or coming to understand something both individually and within a social process is more common in contemporary science education.

Linking Product/Process to Traditional/Progressive. Science as content or process can be linked to the two teaching frameworks previously presented. Science as a body of knowledge corresponds to traditional practices of teaching where transmission of information, in a large group setting, is dominant. Science as a process can be linked to a progressive style where the teacher sets up the learning environment and allows students to investigate and discover on their own. In reality, however, science instruction needs to provide for teaching the process of inquiry as well as for the acquisition of a body of knowledge. Unfortunately, the two pedagogical frameworks so far presented are inadequate for the task.

Science is . . .

". . . scary, but probably the most hands-on subject other than PE."

". . . a complex subject for me . . ."

". . . is boring, but interesting (I like the lab activities)."

". . . often a lot of memorizing
 • experiments that don't seem to have a connected purpose
 • for the smart kids."

". . . hard to comprehend and boring (sometimes)
 • horrible
 • could be exciting or interesting
 • one of my worst fears."

". . . Measurement, Chemistry, Biology, Pollution, Gas, Experiments."

". . . is hard. There are a lot of words to memorize that never seem to make any sense."

Figure 4. Responses to the writing prompt "Science is . . . " made on the first day of class by preservice elementary education students

For example, traditional practices of teaching science neglect the process of science and often neglect the role of students' prior understanding in the construction of new meaning. Also ignored is the function of the social and collaborative nature of meaning-making. Progressive practices, however, often neglect scientific knowledge, which is socially and culturally determined. Such a body of knowledge is pervasive throughout the culture and is transmitted via a variety of media, including text in magazines and newspapers, electronically, through radio, television, and computers, and through oral formats of presentations and discussions. Children cannot "discover" or "construct" all the knowledge that has come before them. Thus, in the ideal of Jefferson (Lee, 1961), an educated populace is one that not only knows how to learn, but also, at the very least, understands the culturally determined body of knowledge that is necessary for active participation within the current society.

In terms of educational practice and the teacher's role within the classroom, it seems useful to construct a framework for teaching that allows for the combination of traditional components, where direct instruction of content may be necessary, as well as progressive practices, where students do partake in open-ended "discovery" activities. Doing so would satisfy both notions of what science is in terms of process and content and what science teaching is in terms of traditional and progressive approaches.

Bridging the Gap Between Traditional and Progressive

Redefining the Teacher's Role in Learning

John Dewey. In The *Child and Curriculum* (1902/1990), Dewey is explicit in defining the teacher's role: "Guidance is not external imposition. It is freeing the life-process for its own most adequate fulfillment" (p. 195, emphasis his). He comments that

> There are those [teachers] who see no alternative between forcing the child from without [transmission of information], or leaving him entirely alone [a discovery approach]. Seeing no alternative, some choose one mode, some another. Both fall into the same fundamental error. Both fail to see that development is a definite process, having its own law which can be fulfilled only when adequate and normal conditions are provided. (p. 195)

As Dewey implies, "adequate conditions" are the combination of guiding and providing proper stimuli, while at the same time allowing for individual exploration, which is crucial in the learning process of the child. As previously presented, Dewey later criticized the progressive movement and reiterated his belief that the teacher has a central role in selecting and implementing those experiences that ". . . do not repel the student, but rather engage his activities . . . " (1938, p. 27), and that traditional teaching represents ". . . not the absence of experiences, but their defective and wrong character—wrong and defective from the standpoint of connection with further experience" (p. 27). In affirming the teacher's role in this process he commented that

> A primary responsibility of educators is that they not only be aware of the general principle of the shaping of actual experience by environing conditions, but that they also recognize in the concrete what surroundings are conducive to having experiences that lead to growth (p. 40).

In Dewey's view, and one supported here, is that the role of the teacher, as the adult and more experienced person in the classroom, must, at times, be a direct one. In this sense, the teacher's direct guidance or explicit teaching is not an imposition to the child's construction of meaning, but an aid and a model for overcoming barriers to learning. Unfortunately, in contemporary science education, teachers view their roles as "facilitators," in which they set up the learning environments but have little interaction with the students thereafter. Dewey (1938) described similar situations, ones we can still see occurring in today's classroom.

> Sometimes teachers seem to be afraid even to make suggestions to the members of a group as to what they should do. I have heard of cases in which children are surrounded with objects and materials and then left entirely to themselves, the teacher being loath to suggest even what might be done with the materials lest freedom be infringed upon. . . . what is more important is that the suggestion upon which the pupils act must in any case come from somewhere. It is impossible to understand why a suggestion from one who has a larger experience and a wider horizon should not be at least as valid as a suggestion arising from some more or less accidental source (p. 71).

I place emphasis on Dewey's statements because too often he is misinterpreted as advocating a strictly child-centered, progressive approach to learning. Upon closer reading, it is apparent that while he favored learning that actively involved students within the process of education and the application of their knowledge to new experiences, he maintained that the role of the educator was crucial in selecting and guiding the students through the process of learning.

Lev Vygotsky. Vygotsky (1978, 1989) also maintained that the teacher's role should be active as she helps students in the learning process through arranging access to new experiences, guiding discussions, presenting information, and facilitating the organization of students' conceptual understanding. Vygotsky realized that the behaviorist model of learning "is impossible and fruitless" (Vygotsky, 1989, p. 150) and was counter to concept formation as a continuously evolving process. He did suggest, as did Dewey, that "interference," or a more direct teacher role may favorably help students in the development of their own concepts. In an attempt to balance apparently opposing ideas, he comments that "Deliberate introduction of new concepts does not preclude spontaneous development, but rather charts the new paths for it" (p. 152).

Vygotsky's (1978) framework is based on two important philosophical perspectives. The first is that there are two types of development that are important in a child's learning. One is the child's actual development and refers to the developmental level that the child is at and in which he can carry out certain tasks independent of outside help or guidance. The other is the child's potential development, the developmental level that the child may attain with the help and guidance of an adult or more experienced peer. Between these two is the Zone of Proximal Development of that child and indicates

> ... the distance between the actual developmental level as determined by independent problem solving and the level of potential development as determined through problem solving under adult guidance or in collaboration with more capable peers (p. 86).

As students close the gap between actual and potential development, it is the teacher's role to continue to structure the learning toward each student's greatest potential. Thus, teacher guidance and help, as well as her decisions about what to teach and when, are necessary within the educational process of the child. Indeed, the sign of good teaching is eventually to have students working independently of teacher assistance and guidance on tasks that were learned early in the teacher/student relationship (i.e., the beginning of the school year). The teacher would then move on to more sophisticated learning, where her guidance would be necessary as the students close the gap between their "new" actual development and the "new" potential development introduced by the teacher.

The other perspective posed by Vygotsky is the assumption that society and culture are central to any learning. Since languages are socially constructed in society, learning, through language use, is a social process. Vygotsky (1978) states,

> Thought development is determined by language, i.e., by the linguistic tools of thought and by the socio-cultural experience of the child. . . . . The child's intellectual growth is contingent on his mastering of the social means of thought, that is, language. (p. 94)

This viewpoint clearly indicates a role of social collaboration among individuals as an important component in learning and the construction of understanding. "Human learning presupposes a specific social nature and process by which children grow into the intellectual life of those around them" (Vygotsky, 1978, p. 88).

Based on these perspectives, the teacher balances her role in helping her students learn between what her role would be in traditional and progressive practices. While direct teaching may be "fruitless," how a teacher engages her students through conversation, questioning, or direct instruction will be crucial as she guides them toward potential development. In addition, as she sets up experiences that are open-ended, how students move from actual to potential development will be dependent upon the teacher's role in the social process.

Cognitive Apprenticeship: Theory into Practice

In trying to redefine the role of the teacher in helping students learn science, one finds it useful not only to borrow from both Dewey and Vygotsky but to view how other theorists have applied these ideas in conceptualizing what the teacher does in the classroom. In aspects of "Cognitive Apprenticeship," Roth (1991,1993) utilizes a notion of the "zone of proximal development," emphasizing that development of the child is dependent upon social interactions with adults or peers. In the words of Rogoff and Lave (1984) "The teaching-learning situation, then, becomes one were [sic] a practicing member of the culture [the teacher] models the use of conceptual and practical tools" (p. 3).

Central to both the formal school structure and the "informal" out-of-school learning is the importance of the relationship between the activity and the context in which it is performed, especially the "social milieu" (Rogoff, 1984, p. 2) in which the activity is conducted. Directly associated with the social milieu is the use of culturally determined tools, sign systems, and practices that "facilitate reaching appropriate solutions to problems" (Rogoff, 1984, p. 4). Directly linked to this is the idea that the use of tools and practices are ". . . transmitted to children and other novices through interaction with more experienced members of society" (p. 4). The social system, therefore, acts as a funnel for cognitive development and this development (learning) "is guided by social interaction to adapt to the intellectual tools and skills of the culture" (Rogoff, p. 4).

Within a cognitive apprentice viewpoint, teacher guidance acts to "enculturate" the learner into the trade, system, or "community" of which the more knowledgeable person is a member. It is the teacher's role, as the "expert" member of the learning community, to help students gain experience with the practices of science (questioning, investigating, and solving problems), as well as with the appropriate language of the field. Consequently, the implicit knowledge of the practitioner's practice (the teacher) is made explicit to the apprentice (the student). It is, therefore, the adult's responsibility to arrange appropriate tasks, to facilitate the learning by monitoring and regulating task difficulty, and to model not only the actual performance of task completion, but also that of the social processes embedded with the learning.

As a framework for science teaching and learning, the middle ground is not teacher-centered or child-centered, but a combination of the two, where both teacher and student have roles and responsibilities within the learning process. When suitable, the teacher may take on a more traditional role, where she openly influences the direction and the connections the students may make. She may guide and model mature behavior and skills that are essential to the learning process. At other times, she allows for investigations that are student directed and student organized, especially when the students are practicing the skills and behaviors that have been previously modeled by the teacher.

Therefore, it seems clear that the role of the teacher is to move constantly in a dynamic balance between traditional and progressive teaching frameworks. The purpose of this balancing is to design learning experiences that also move back and forth between the actual and potential development of the students. The goal is to guide the students to their maximum potential, moving toward their "new" actual capabilities, only to challenge them with new experiences that advance their potential even further.

Tying It All Together: Moving Toward the Middle and a New Constructivist Perspective

As a "mantra-like slogan" (Cobb, 1994a, p. 4) where students individually construct their own knowledge, constructivism is limited as a means to explain the role of a teacher in the classroom. An individual approach to learning is an interpretation of constructivism that places it firmly in the progressive framework. However, in the recent past the theory of constructivism has been described as having an individual and a social component, derived from Vygotsky's perspective on learning and development (Cobb, 1994a, 1994b; Driver et al., 1994; Tobin & Tippins, 1993). In this sense learning science is an individual and at the same time a social process and "involves young people entering into a different way of thinking about and explaining the natural world; becoming socialized to a greater or lesser extent into the practices of the scientific community . . ."(Driver et al., 1994; p. 8). Learning in science, therefore, moves beyond "personal enquiry" and "discovery learning" and engages students socially, where the teacher is ultimately responsible for modeling those skills deemed necessary for continued learning.

Reconsidering constructivism as only a "referent" for teaching and learning (Tobin & Tippins, 1993) and not a prescription necessitates a reconsideration of the teacher's role in science education. Earlier notions of constructivism had the teacher setting up the learning environment, only to "facilitate" student learning. However, facilitate, as a teaching process, has multiple meanings and essentially can be, and has been, interpreted as implying a "hands-off" approach to learning. The argument here is that the role of the teacher is greater than that of just a facilitator and becomes more clearly defined as allowing for more direct teacher interaction with the students as they struggle to construct their own understanding of science.

Further Thoughts

As Peshkin (1993) stated, when we understand the processes of classroom life ". . . we understand something of value" (p. 24). Therefore, if teachers are the crucial link to fundamental reform in curricular change and science teaching practices, then addressing how and what teachers do in the classroom will be helpful in developing a better understanding of their influence upon the enactment science. This chapter has attempted to construct an alternate view of the teacher's role within the classroom, a role that is situated between traditional and progressive aspects of teaching and learning.

The purpose here is to use this single case study, this story, as one example of how a teacher constructs her role in the classroom and how she and her students enact the science curriculum. In order to develop a theory of teaching and learning science, it appears we need to observe and describe fully how teachers, in a variety of contexts, implement innovative science curricula. As Schwab suggested, the goal is not to define a singular framework, one generalized to all situations, but to focus on those practical applications of the teaching process. By doing so, we shall add to our knowledge base of what is useful and what works in practice. In order to change what happens in the classroom, we may find it necessary to alter our perceptions of what it means to teach science.

CHAPTER THREE
The Role of Teacher Beliefs in Enactment

Few would argue that the beliefs teachers hold influence their perceptions and judgments, which, in turn, affect their behavior in the classroom, or that understanding the belief structure of teachers and teacher candidates is essential to improving their professional preparation and teaching practices. (Pajares, 1992, p. 307)

Teacher Beliefs

Pajares's (1992) statement reflects a common conclusion of research on teacher beliefs. Mainly, what teachers believe, with regard to their teaching and learning, will drive their practice (see Clark & Peterson, 1986; Kagan, 1992; Munby, 1982, 1984; Nespor, 1987; Tabachnick, Popkewitz, & Zeichner, 1979). Unfortunately, while inquiry into this topic is central to a complete and useful understanding of the thought processes of teaching, research on teachers' implicit beliefs is considered the smallest and the youngest part of the literature.

In attempting to discover a source of teacher beliefs, Nespor (1987) argues that beliefs are structured from previous events and experiences. A teacher's past events creates "guiding images" that act as a filter for new information. A belief structure created from an earlier experience may also be resilient enough as a standard that newer information is compared to. For example, if a teacher changes conceptions of what quality teaching is, from a traditional whole group approach to a cooperative learning orientation, all new information about practice will be filtered through the cooperative learning belief structure. Or, if, as a child, she had positive experiences with a specific teacher and these experiences influenced her viewpoint of how to deal with students, then this history may motivate her to act in a similar manner and thus treat her students as she once was treated. Based on these ideas, Nespor (1987) suggests that a person's memory serves as a template for future action and that revealing these memories through interviews can provide insight as to the belief structure that guides a teacher's practice.

Pulling from work on teacher metaphors is also helpful in ascertaining how beliefs mold practice (see Tobin, 1993; Tobin & Espinet, 1989; Tobin & Gallagher, 1987; Tobin et al., 1994; Tobin & Ulerick, 1989). Metaphors may be used to help define a teacher's role in the classroom and thus guide her practice. A "teacher as intimidator" (Tobin & Gallagher, 1987) may be one who is authoritative, demanding respect, and squelching any potential misbehavior. A "teacher as preacher" (Tobin

& Espinet, 1989) is one who may use sermon-like lectures as the dominant teaching technique. A teacher who views herself as a resource may then structure the classroom for collaborative and independent work, where students ask for help and guidance when needed. (For a detailed description of metaphors and beliefs, see Tobin, et al., 1994). These studies also suggest that a teacher may use more than one metaphor to fit certain classroom contexts (policeman, mother hen, or an entertainer). Use of metaphors may also aid in teacher change as a teacher attempts to alter traditional practices to those that fit more with contemporary views of science education (Fawcett, 1992; Tobin & Ulerick, 1989).

Teacher beliefs and teacher metaphors, therefore, appear to play a major role in how the teacher structures and organizes the classroom. Thus, looking at how beliefs impact and are manifested through practice will prove useful in understanding how and why teachers teach. In this chapter, similar to the structure of Chapter two, I will use previous research on teacher beliefs in the attempt to construct a theoretical framework that can then be used for a lens for viewing the enactment of science.

Defining Teacher Beliefs

Studying beliefs is problematic because they cannot be directly observed and must be ascertained by what people say and do. A variety of terms are used in reference to teacher beliefs. These include *preconceptions, implicit theories,* or *orientations*. Teacher perspectives, describe by Clark and Peterson (1986), are ". . . a combination of beliefs, intentions, interpretations, and behavior that interact continually" (p. 287). Tabachnick and Zeichner (1984) suggested that these perspectives include not only beliefs about work as a teacher but also how these beliefs are given meaning through behavior and classroom practice. Whatever the definition, it is generally agreed that what teachers believe in, as it relates to their philosophy of teaching, their role within that process, the role and expectations of the students for learning, and the role of the school, science curricula, and context for instruction will be an essential foundation for what occurs in the classroom.

Teacher Efficacy

In attempting to link theories of teacher beliefs to the practice of teaching, Smylie (1990) suggests that "teacher efficacy is believed to be one of the most significant social-psychological factors influencing teachers' work" (p. 48). Discussions of teacher efficacy are most often based on

Chapter Three: The Role of Teacher Beliefs in Enactment

Bandura's ideas of self-efficacy (1977, 1986), which defined this notion as "people's judgments of their capabilities to organize and execute courses of action required to attain designated types of outcomes" (Bandura, 1986, p. 391). A person's self-efficacy, however, does not concern the actual skill level of a person but pertains only to their judgments of how they perceive "... what one can do with whatever skills one possesses" (p. 391). There is also a distinction between *self-efficacy*, a perceived level of capability, and *outcome expectancy*, the judgment that behavior will produce certain outcomes. The two are differentiated as follows:

> ... individuals can believe that a particular course of action will produce certain outcomes [outcome expectancy], but if they entertain serious doubts about whether they can perform [self-efficacy] the necessary activities such information does not influence their behavior. (Bandura, 1977, p. 193)

In addition, studies of self-efficacy, which Pajares (1992) contends is a sub-construct of a person's belief structure, may need to be context specific and "... relevant to the behavior under investigation to be useful to researchers and appropriate for empirical study" (p. 315). While an opposing viewpoint criticizes the fact that too many studies of teacher beliefs are narrowly constructed (Kagan, 1992), others indicate the necessity of studying teacher beliefs that are embedded in a specific context and then linking these beliefs to greater belief structure of the individual (for example, linking a teacher's beliefs of science instruction to her overarching theory of education; see Bandura 1977, 1986; Pajares 1992).

Here, I attempt to unveil both the teacher's general beliefs about teaching and learning and her specific beliefs about science instruction. Observing her actions and classroom practice as it relates to her beliefs in teaching science is an important aspect of this study. However, discerning when and describing why certain practices are employed (traditional, progressive, or middle ground), and relating these to a teacher's beliefs, will continue to be the main focus of this study.

Teacher Efficacy in Science Teaching. In constructing the Science Teaching Efficacy Belief Instrument (STEBI), which is intended to measure both self-efficacy and outcome expectancy of preservice and inservice science teachers, Riggs and Enochs (1990) identify two main dimensions of a teacher's efficacy. One dimension is described as "Teaching Efficacy" and relates to the teacher's outcome expectancy. The other dimension is "Personal Teaching Efficacy," which specifically refers to the teacher's self-efficacy and beliefs concerning her ability to teach science. The intent is to link what a teacher believes about her ability to teach science to her potential for having these beliefs manifested in practices that lead to positive student outcomes. Those teachers who have a high self-efficacy will most likely be confident in their ability to teach science while those who have a high outcome expectancy may believe that their teaching and any innovations used in instruction will be exhibited in high student achievement and interest in science (Riggs & Enochs, 1990).

While scales or instruments may be useful for getting at the generalities of a teacher's beliefs and how these beliefs may relate to classroom practice, we must be cautious about inferring direct relationships between the validity of these measures and the actual teaching of science. Pajares (1992) raises this caution when he asserts that inventories or measures used to ascertain beliefs

> . . .cannot encompass the myriad of contexts under which specific beliefs become attitudes or values that give fruition to intention and behavior. Individual items fall prey to "it depends" thinking, and responses fail to provide either accurate or useful inferences of behavior. (p. 327)

Munby (1984) also cautions against the use of traditional belief instruments because the responses of teachers may be more related to the beliefs of the test developer, who chooses the items, than to the actual beliefs of the teacher. Thus, in order to lend authority to the results of belief instruments, we can use interpretive research methods, such as ethnographies, case studies or oral histories to provide detail as to how beliefs may, or may not, be manifested in classroom practice. These studies can also account for the unique character of individual teachers within their specific settings. Ultimately, these findings may help us link classroom practice more closely to student outcomes and assist us in gaining detailed insight into the ". . . beliefs that are consistent with effective teaching practices and student cognitive and affective growth" (Pajares, 1992, p. 328).

Linking Teacher Beliefs to Curriculum and Context

Bandura's (1986) self-efficacy theory can also be applied to how the teacher perceives her ability to implement curricula and to do so within the specific classroom context. Bandura (1986) states that while persons may have the skills for performing certain behaviors, "disincentives and performance constraints" (p. 395) may prevent them from doing so. Lack of support from others (teachers, parents, or school administrators), or a lack of suitable equipment or supplies will, not allow a person to perform adequately an intended behavior (i.e., teaching).

> Efficacious artisans and athletes cannot perform well with faulty equipment, and efficacious executives cannot put their talents to good use if they lack adequate financial and material resources. Physical or social constraints further impose limits on what people can do in particular situations. (Bandura, 1986, pp. 395-396)

Physical and social constraints, therefore, can impede a person's ability to perform, and in doing so, the self-efficacy may be higher than application allows. It is this concern of outside resources influencing a person's ability to perform a job that necessitates a brief look at how the curriculum and the context may affect teacher beliefs.

The Curriculum

If what a teacher believes in strongly influences what she does in the classroom, then her beliefs about a specific curriculum will influence its implementation. Changes to a curriculum may occur in "pre-active" planning or during the actual implementation process (Clark & Peterson, 1986). How teachers

> ... adopt, adapt, or reject new curriculum and instructional strategies [depends] in large part on their conceptions of academic subject matter and theories of teaching that subject matter (Smylie, 1992, p. 14).

Thus, there is a strong connection between a teacher's beliefs and classroom practice and curriculum implementation (see Hawthorne, 1992; Munby, 1984; Olson, 1981).

In an attempt to discern the influence that a new science curriculum had on teacher practice, Olson (1981) revealed that the most important element in the teachers' implicit theories was that of classroom control. Teachers viewed their role in the classroom to be one of high teacher control over the structure and orientation of the curriculum. What was found was that regardless of what the curriculum authors intended, the teachers rearranged the curriculum to fit their beliefs of how to structure the learning. Referred to as "domesticating" the curriculum, teachers changed discussions to lectures, skill development activities became memorization tasks, integration was turned into a patchwork of topics to be pieced together, and performance tasks often became traditional multiple-choice type tests. After an initial experimentation period, teachers would re-exert their influence over the curriculum and how it was implemented. The outcome is that the original intentions of the authors were frequently undermined.

In another study Hawthorne (1992) asserted that teachers perform a "balancing act" between their professional obligations to the system and to the students and to their desire to maintain their autonomy as independent thinkers as they decide what is best for the children. Teacher freedom to make professional decisions ". . . lies at the heart of the reform debate and at the heart of . . . teachers' curricular decision making" (p. 128). Typically, curricular decisions made by teachers are localized or "context-bound," and teacher decisions relate specifically to the students and to the needs and demands of the specific classroom.

Based on these studies and others like them, one conclusion is that the focus of educational reform may best be centered on the classroom, incorporating the teacher and the students and the decisions teachers make in the enactment process of curricular implementation. As presented, these decisions are based on teachers' beliefs and judgments as they attempt to satisfy the many obligations placed upon their profession. Similar to Hawthorne's focus, the aim of this study was to discern the influence that the science curriculum had on the teacher and her decisions in the classroom, as well as to learn how the teacher may have changed and adapted the curriculum to fit her needs.

The Classroom Context

The classroom context in which the teacher and students operate also influences the decisions that teachers make during curriculum implementation (Oakes, 1985, 1990; Sizer, 1985). Sizer portrays a classic compromise that Horace, a fictitious high school English teacher, must make between the reality of what he wants to do and what he can feasibly do with his students. Constraints relating to the classroom context, overcrowding, curricular mandates, and teacher time in the classroom all influence the decisions a teacher makes in regard to her practice. These contextual components may be many and can include teacher beliefs, experiences, knowledge, expectations, and attitudes toward teaching and learning. There are also the students and their belief systems, their perceived ability and their conception of the usefulness of the curriculum, parental support for achieving success, and the students' socioeconomic background. The availability of school resources, class size, and a genuine climate of support for both the teacher and the students also greatly impact teaching and learning. The principal, the key leader of a school, plays a vital role in supporting the teacher in implementing new practices, and the community's involvement either through funding or by parental means, may influence what happens in the classroom.

A variety of factors then, may influence how science is taught. These include, but are not limited to educational change at the institutional level, the influence of the specific classroom environment on student learning and their attitudes toward science, the influence the environment has on teacher satisfaction and effectiveness as well as student access to high quality learning experiences, teacher expectations, and instructional types. Referred to as the "classroom press" (Huberman, 1983, cited in Fullan, 1991), these factors can exert a tremendous influence on teachers and instruction.

Tracking students by academic ability greatly influences how teachers teach, how students perceive their ability to succeed, and the quality of instruction and materials they receive (Oakes, 1985). The effects of tracking on science instruction, particularly for females and ethnic minorities, revealed the same patterns as in her earlier study (Oakes, 1990). Low-track students had fewer opportunities and less access to science learning experiences that were of the high quality usually experienced by high-track students. Low-track students also had science teachers who had less experience and lower expectations in their students' ability to succeed.

In contrast, higher-track classes often were characterized by higher teacher expectations with learning situations that were associated with challenging and relevant curricula that emphasized higher-order skills. Students in high-track classes also received greater access to the most challenging and advanced courses and thus usually got the highest quality instruction and experiences (Oakes, 1985, 1990).

This work by Oakes is relevant here because it examines a direct classroom effect of one factor that may influence teaching and learning and brings us full circle back to the original discussion by emphasizing that teacher beliefs may be the most important contextual feature affecting curriculum implementation. In addition, Oakes (1985, 1990) and Hawthorne (1992) provide evidence that looking at the specific classroom context may be most informative in understanding the influences upon a teacher's practice.

Accordingly, this study does not examine educational change and does not address curriculum implementation at the institutional or even district level. What this study focuses on is curriculum implementation in one urban classroom, at the level of the teacher and the child, explicitly describing what influences there may be that influence the enactment of science curriculum.

Final Thoughts

> Teachers and their students stand at the center of educational practice and thus represent the focal point for educational reform. Teacher efforts to create and sustain improvement within the classroom coupled with necessary support and leadership from school and district organizations represent our best hope for effectively managing the creation of classroom curriculum and improving educational outcomes. (Hawthorne, 1992, p. 131)

In order to address the need for changes in how science is taught we must first understand how and why teachers do what they do in actual classrooms. Hawthorne (1992) argues that the focus of change should center on the teacher and her students. Within science teaching, Yager and Lutz (1994) stress that the emphasis should be on the "how" of classroom practice as oppose to the "what" of the curriculum. Sizer (1985) stresses the importance of local adaptations that meet the needs of the students within a specific context, and Hawthorne (1992) asserts that such "personalization of the teaching process occurs not at the level of the district or school organization but at the level of individual classrooms" (p. 129).

In studying how one teacher enacts science in the classroom it becomes crucial to consider her beliefs and impressions of her students and of the school within the localized context of her practice. Studies of exemplary teachers have generally found that they focus on ". . . learning with understanding, use strategies to encourage students to engage in higher-level cognitive tasks and maintain a classroom environment conducive for learning" (Tobin, Kahle, & Fraser, 1990, pp. 3-4). Such teachers promote the development of inquiry skills, conceptual understanding, and the creation of positive personal attitudes.

Exemplary teachers, therefore, regardless of track level or perceived ability of the students, have the belief that all children can learn and can benefit from quality classroom experiences and, thus, if allowed, adapt instruction to the levels of their students. Positive beliefs, a sense of efficacy, and the support she receives from her principal and her colleagues will all influence what and how she teaches. Counter to many of the studies previously cited, what this study attempts to show is one example of what is "good" about science teaching, regardless of student background and how a teacher engages students in the learning process.

CHAPTER FOUR
Getting "at" Enactment:
Constructing a Method of Inquiry

> Any biologist who observes a tiger, gorilla, panda, or other creature and says he or she has done so with total objectivity is ignorant, dishonest, or foolish. (George Schaller, *The Last Panda*, 1993, p. 105)

Study Design

The Problem of Finding a Teacher

The intent of this study was to investigate the curricular enactment of one exemplary teacher as she implemented innovative science curricula. The focal point was on what went well and what was good about teaching instead of continuously focusing on what's bad in education. An additional perspective was that much of the literature in science education has either addressed curricular issues and proposed methods of teaching or student outcomes as reflected in the analysis of achievement tests or attitudinal scales. There appeared to be only a few studies that were looking directly at what teachers actually did in the classroom. This lack of research seemed especially true if the practices were good and teachers were having success.

I used an ethnographic case study method for two reasons. First, the study required no control over behavioral events, and addressed contemporary classroom issues. Thus, the investigation fit under the umbrella of a case study (See Yin, 1994, p. 6). Second, and because I fully intended to observe and collect data throughout the entire school year and I was interested in how the teacher helped to create a classroom environment conducive for learning science, I was attaining a "cultural description" (Wolcott, 1975, p.112). Therefore, the study also fell under the realm of ethnography (Spradley, 1980; Wolcott, 1975).

My initial and continuing question was based on the "how" of a teacher's practice. With this goal in mind, and in trying to find a teacher and a classroom, I decided to focus on the middle grade levels, anywhere from grades five to seven. I based this decision on the fact that these grades were a little closer to my experience with ninth graders (even though I had worked with second graders), and this closeness would be important as I began to make links between what was going on in the middle grades to what goes on in high school. In addition, I could move backwards, to the lower grades, using ideas I saw here to help younger students learn. I also felt more comfortable with older students, and if I ever wanted to incorporate student discourse into this study, I believed that the older student-talk would be richer (which, of course, is not necessarily true). I focused on a self-contained classroom because whether it is at the elementary or middle level, this is one area of teaching science that needed more investigation. Finally, I wanted an urban classroom, because many urban students get fewer opportunities to engage in meaningful science learning, and thus if worthwhile instruction was going on, it would be worth looking at.

Finding the type of teacher was not a random process, and in actuality, it was the teacher more than the classroom that I was looking for. I began a year ahead of time by asking colleagues and friends if they knew of any teachers who, in their opinion, were doing an outstanding job in teaching science. Not wanting to impose my opinion of what "good" meant, I never fully expanded on this notion because my idea may not have been the same as others. Thus, I simply relied on the opinions of those I asked. However, because those people I asked were fellow graduate students or those involved in innovative math/science programs, the idea of "good" seemed implicit. Ideally, I was looking for someone who was a "doer," someone who knew the importance of getting the students involved in learning, who was willing to experiment with alternative methods, and who believed that science was an important part of her students' education. I was also looking for someone who was doing science consistently throughout the school year. Random activities and haphazard lessons would not get at the heart of my study. Essentially, I was looking for someone who was having "success" in teaching science, however it was measured.

I had the usual problems with entry into classrooms. These included teachers not returning my initial phone calls or asking me to wait a few weeks (or even a few months) so I could see a particular topic or a specific pet lesson. I even had one teacher who had agreed to let me observe for a few weeks beginning after Christmas break, only to cancel at the last minute. Finding a classroom where science was consistently being taught and finding one where the teacher was confident enough that I was not evaluating her teaching turned out to be more of an issue than I had first considered.

It took from September to March before I found Donna. She was recommended by a student of one of my colleagues as being a teacher who was doing "great" things in the classroom, especially in the area of cooperative learning. She was considered innovative and tireless. I was told that the kids loved her and the parents appreciated her. After an initial meeting and one observation, we sat down, and I asked her if she would mind if I spent the following academic year observing her classroom and investigating how she went about teaching science. I explained to her that it was nonevaluative (she didn't "care" if it was), that she would be the center of my inquiry (she is very protective of her kids), and that it was for my dissertation. She agreed, and as they say, the rest is history.

The School and Classroom Setting

The study took place at the Oppenheimer Elementary School, an urban elementary school on the northwest side of a large Midwestern city. This school serves children from a wide variety of ethnic backgrounds; African American, Anglo-American, Latino (including Cuban, Puerto Rican, Mexican), and recent immigrants from Ukraine and the former Soviet Union. The school's largest population consists of Latinos, mostly from Puerto Rico, and about one third of the children were from Mexico. Approximately 70 percent of the children qualified for the state subsidized lunch program, giving evidence to the low socioeconomic status of many of the students.

Oppenheimer is considered a magnet school, consisting of *Maintenance Bilingual, Whole Language*, and math/science programs. Students are chosen by lottery and not by academic performance. A number of children who attend Oppenheimer have parents who work there as faculty, staff, or aides. It is a nontransient school population, with up to 97 percent of those students who enter in kindergarten going on to graduate from the eighth grade (see Chapter six for more detail).

Getting "at" the How of Teaching

Hawthorne (1992) makes the following point as she describes the data collection strategies in her study.

> Educational criticism, as a form of inquiry, does not dictate any one set of data collection and analysis. Instead, it allows the critic freedom to tailor in, an idiosyncratic manner, methods of data collection and analysis to the phenomenon under investigation. (p. 136)

I find solace in this statement because I too did not follow one recipe for data collection and analysis. However, while I was eclectic in my methods, I did stay close to those recommendations and examples set forth by seasoned veterans of the task (see Eisner, 1991; Erickson, 1984; Glesne & Peshkin, 1992; Peshkin, 1993; and especially Wolcott, 1975,

1990a, 1990b, 1994). How I approached my writing was again influenced by Wolcott (1990b, 1994), as well as by Norman Maclean's *Young Men and Fire*, the spectacular and horrific account of the 1949 Mann Gulch fire; Peter Matthiessen's *The Snow Leopard*; and George Schaller's *The Last Panda*. Specific to method, however, my approach most closely followed the one outlined by Spradley (1980) in his book *Participant Observation*.

The General Process. Data collection consisted of three main phases, which Spradley (1980) likens to a "funnel." During the first phase, the most general component and widest point of the funnel, I was a passive participant observer. It was here that I collected, at least two days a week, *descriptive* data on every aspect of the classroom life as it pertained to the teacher and to her practice. I began the day in language arts and continued through the other disciplines to the end of the day, when the class "did" science. In all, I observed about twenty-seven-and–one-half hours of science instruction, approximately sixteen hours of math science labs, and close to six hours of science fair (Table 1).

Number of Observed Classroom Lessons and Associated Times

	Language Arts	Math	"Labs"	Science	Science Fair	Social Studies	Total
Number of Sessions	5	5	16	30	4*	13**	73
Number of Hours	4.94	3.38	16.24	27.53	5.84*	13.07**	71

*One session and three hours were for science fair judging.
**One session and five hours were for a social studies field trip.

Table 1. Number of Observed Classroom Lessons and Associated Times Language

During my initial interviews, I collected general information regarding Donna's background, her teaching beliefs, her beliefs regarding teaching science, and her expectations for the students. Phase one lasted from September (the beginning of the school year) through most of November.

Data analysis commenced from the onset. Based on the early analysis, the second phase, beginning toward the end of November, consisted

Chapter Four: Getting "at" Enactment

of *focused* observations and interviews. I began to concentrate more specifically on science teaching and science-related activities. Additional questions and themes began to emerge, which I addressed either through informal discussions with the teacher or by more structured interviews. It was during this phase that my role moved toward a "moderate participant observer," attaining active participant status on a limited basis, during one field trip and two outdoor science activities. Phase two lasted through science fair, which was toward the end of February.

The third phase began near the end of the school year. At this time I began to get more *selective* with my observations and concentrated on particular science lessons related to specific emerging themes. These observations culminated in June with the "oil spill" activity, which consisted of six consecutive daily lessons.

The final interview focused information regarding specific issues and questions that occurred throughout the year. I used this opportunity to revisit previous topics to confirm earlier findings. The entire process proceeded from the general to the specific, narrowing as the year moved on. (see Appendix A for specific data collection).

What Is It That I Have? Data Analysis.

The General Process. Following an approach similar to Spradley's Developmental Research Sequence, I began looking at my *descriptive* observations. Once I identified the commonalities of the classroom flow, I then reviewed the data to look for emerging patterns and themes of teacher behavior as she taught her lessons. In my "domain analysis," I identified major features. One feature, for example, was her heavy emphasis on training students in the use of social skills. The emphasis was so strong that it actually could be considered part of the science content for this class. (This finding represents a strong link between teacher beliefs and curriculum) I also recognized instances of teacher modeling, which was another emerging theme of the study.

Once having identified general domains and having conducted *focused* observations and interviews, I identified more instances that supported the emerging themes and how these influenced classroom practice. In addition, my theoretical framework was imperative to my analysis of the teacher's actions and words. I compared the collected data to my initial framework, looking for a match between data and theory. Termed "pattern matching" (Yin, 1994), I was looking for potential confirming and disconfirming evidence that represented a "middle-ground" between traditional and progressive approaches to teaching science.

Initially, the model I had proposed earlier did not come directly into play. In later analysis, however, I began to use my descriptive observations of enactment to "backward map" (Elmore, 1979-1980) from classroom practice to the other pieces of the model (teacher beliefs, curriculum, and context). I was looking to identify those model components that may have a specific influence upon the enactment of science.

Accuracy, Comprehensiveness and Subjectivity

There comes a time in all studies where one has to assure the reader that what is presented is what really happened. In the traditional empirical research approach one must address the strict statistical definitions of reliability and validity. However, these terms take on new meaning or new emphasis in qualitative research. Thus, I favored the techniques of those qualitative researchers who have defined and described how these issues relate to their practice (see Bogdan & Biklen, 1992; Eisner, 1991; Miles & Huberman, 1994; Wolcott, 1990a, 1990b).

In a qualitative study such as this one reliability means accuracy and comprehensiveness (Bogdan & Biklen, 1992). Accuracy relates specifically to the fit between the recorded data and the actual observations and events within the setting. If the descriptions are not accurate, the data are unreliable. Comprehensiveness means that there are numerous sources of data that support the descriptions, including direct observations, interviews, tape recordings of conversations, and video. Each data source could then be used to provide a clear and comprehensive picture of the study, which would then support its accuracy. Termed "structural corroboration" by Eisner (1991), accuracy and comprehensiveness, therefore, mean that multiple data sources are needed to lend either support to, or contradict, interpretations and evaluations. Regardless of the type of research one does, the main idea is to look for "evidence that breeds credibility" (p. 110), which then supports the evidence's reliability.

Taking a different view of reliability and validity does not mean that as a qualitative researcher I accept everything, or that "anything goes." As a scientist, I attempted with great care to present a case as detailed and as accurately as possible. I may not get it all right, but I have gone to "considerable pains not to get it all wrong" either (Wolcott, 1990a, p. 127). While you the reader may be able to judge from my previous description whether my methods were "good," I will present here what I did to address the issues of accuracy, comprehensiveness, and subjectivity.

Subjectivity

I began this chapter with an epigraph by George Schaller (1993), arguably the most eminent animal behaviorist of our time, who stated that any scientist who claims total objectivity is "ignorant, dishonest, or foolish." In the introduction to this book I presented my potential biases toward this study because I wanted my subjective influences to be known early on. However, as Schaller implies, subjectivity need not be a liability.

Educational research allows us, through our own personal experiences, to provide insight into the specific context of the classroom and to make use of subjectivity in a positive manner. "Widespread familiarity with classroom life provides a reservoir of experience upon which most can draw to determine the empirical adequacy of the critic's portrayal" (Hawthorne, 1992, p. 141). Thus, the author as well as the audience of a qualitative study can use experiences in schools for gaining access from an insider's perspective. Such a perspective may, therefore, allow the descriptions and stories to "ring true to the reader's own experience" (p. 141). Referred to as *referential adequacy* by Eisner, my ability to connect with my readers (partially a result of my own experiences as a teacher), and to allow them to see a portion of classroom life that may have previously gone unnoticed, increases the validity of the study.

Of course my familiarity with classrooms may also streamline my focus of what the teacher does and blind my observations and interpretations to what I wanted to see. I again owe a debt to Schaller who suggests that all research is subjective in nature because we all make decisions about what to study and what data to collect. In light of this we still must attend to the "elusive criteria like balance, fairness, completeness, sensitivity" (p. 133), or what Wolcott calls, *rigorous subjectivity*.

Accuracy

There were numerous ways by which I made sure that what I got was accurate. Initially, as a passive observer I talked little, listened a lot, and recorded as accurately as possible the activities in the classroom. Using audiotaped recordings of classroom sessions, I was able to obtain precisely, in most instances, what Donna was saying and how she used talk to organize her class. I also enlisted her feedback on numerous pieces of data, including the description of her professional history, her description of the origin of labs, all but one formal interview, and a draft form of "Curriculum," "Context," and "Teacher Beliefs." I also discussed with

Donna the emerging themes of the study. Each interaction allowed her to confirm, elaborate, and provide input as to the accuracy of these themes.

In Chapters seven and eight I have attempted to allow the teacher and the classroom descriptions to "speak for themselves." By doing so I have attempted to provide you with a direct view into what transpired in the classroom. However, with all great plans come compromises. In Chapter eight the primary data of teacher talk is limited mainly to the section describing the BSCS curriculum and science fair. The other component, labs, has a reduced amount of direct teacher talk, and thus is not presented chronologically. This is due to the structural nature of the labs where there is less teacher talk. Science fair also has reduced talk, and this is in large part because of the reduced amount of observation time.

Comprehensiveness

Triangulation" and "structural corroboration" are achieved through the use of observational data, audiotape recordings of classroom practice, formal and informal interviews, school and classroom documents, and discussions with secondary sources. Data from each source are used to bolster the data from the other sources. Again, the idea is to attain enough data, each supporting the other, so that the weight of the evidence is compelling enough to convince the reader.

Miles and Huberman (1994) indicate that another way to "get it right," and to counter the effects of bias, is for the researcher to be immersed in the field for an extended period of time. By observing at least two days a week for most of the school year (and occasionally three days per week), and by following up in June with the "oil spill" session, an activity that extended for six consecutive days, I specifically made sure I "got it right." Thus, my extended time in the field insured both the study's accuracy and comprehensiveness.

In addressing the issues of accuracy, comprehensiveness, and subjectivity, I made every attempt to increase the credibility of my findings. By taking a descriptive narrative approach to writing, I allow the reader numerous opportunities to make connections between her own experiences and the ones portrayed here, as she judges the study's worth for herself.

Final Thoughts

When I read Schaller's *The Last Panda*, I was not only intrigued by the masterful portrayal of the life of one of the world's magnificent creatures, but I was also saddened by the ultimate conclusion that this animal may

Chapter Four: Getting "at" Enactment

soon vanish from the earth. For me, Schaller "got it right," but in reading this piece I realized that while a beautiful story was unfolding, behind the scene were endless hours of painstaking observations, data collection, chemical examination, and statistical analysis. Schaller himself admits this to be true but steers us toward his technical report if we so wish to know exactly how many dung samples he collected and analyzed.

In his "story" Schaller conveys the human side of research and the importance of recognizing and maintaining a distinction between the technical and humanistic characteristics of science and emphasizes that as humans, we do impose our own selves upon the study, and in doing so we bring a certain richness and reality that is not apparent in a more formal presentation of scientific data. He states:

> Lacking in such reports is the human factor, the joy of discovery, the pleasure of new insight, the admission that research is sporadic and haphazard—and the fact that the information is not as objective as one would like to think. Statistics may help to describe the universe but not other beings: numbers cannot convey the quality of a creature, they cannot express love, anger, joy, and courage ... A fact is not a fact until someone has posed the question, and slowly the world of an animal emerges from the questions raised and facts collected. But if someone else asks a different question, a different creature, a different reality, comes into existence. The animal is an illusion created out of the animal's interaction with an observer who decides what to measure and record and what to ignore. We constantly infer the unseen, we confuse ideas with facts. (Schaller, 1993, p. 105)

Understanding everything that goes on in this classroom is an enormous task, and even comprehending and documenting all of what goes on during science instruction proved challenging. Here, a concerted effort was placed on attaining information about Donna's beliefs about teaching science, documenting her classroom practices, and attempting to establish a link between beliefs and practice, especially as these may relate to how she plans and teaches. Is there a categorical style of teaching: Traditional or Progressive? Or is her style more eclectic and pragmatic, a middle ground that fits her needs and those of her students, including what she believes to be important for the lifelong learning of these children? Answering such questions, or at the very least gaining insight into what she does, may prove useful in our understanding about how one teacher teaches science.

CHAPTER FIVE
The Science Curriculum

The Main Pieces of the Puzzle

There are three main science components in this classroom: *Science for Life and Living* (BSCS, 1992a, 1992b), math/science labs, and science fair. Of the three, the science fair takes up the least amount of time, about a month-and-a-half for preparation and presentation, with math/science labs and BSCS making up the bulk of the implemented science curriculum. "Labs," as they are called, are conducted one hour every week (usually from 12:30-1:30 on Wednesdays), for the entire school year, and the BSCS curriculum utilizes the most time, with four units extending about one to one-and-a- half months each.

The order of discussion will be as follows.

1. The BSCS curriculum and the main components that impact teaching,
2. The math/science "labs" that are teacher designed and implemented and,
3. The science fair.

BSCS can be grouped into two main categories; the *Instructional Material*, the "what" of teaching: and the *Classroom Structure*, the "how" of teaching. There is, of course, an overlap between these categories and this matter will be brought out in more detail in this chapter. Math/science labs were created by Donna for integrating the NCTM standards into student learning. The structure focused on collaborative learning with an inquiry orientation for the integration of mathematics and science. Science fair was similar to the BSCS curriculum with an emphasis on organizational and instructional issues. Unlike labs and BSCS, science fair was the only curricular element that was disliked by Donna.

Science for Life and Living: BSCS

The instructional material of BSCS contains the following elements.

1. The *constructivist* "*Learning Cycle*" that uses the Five "E's," (engage, explore, explain, elaborate, and evaluate. BSCS, 1992b, p. T26).
2. An integration of science, technology, and health,
3. A year-long focus centered on a major *theme, Balance and Decisions*, which is then broken down into one major concept (Balance) and one major *skill (Decisions)* and, a set of sub-skills for each unit, with *Constraints and Trade-Offs* being the sub-skills of the sixth grade technology unit.

The Classroom Structure of BSCS, which has a direct influence on instruction, is as follows:

1. A built-in *cooperative learning model*, placing a distinct emphasis on classroom management mechanisms,
2. The use of *team skills* as an organizing feature,
3. An emphasis on the *inclusiveness* of all students, especially females, disabled students, and those from ethnically and culturally diverse backgrounds,
4. An emphasis on *hands-on/minds-on* activities, and
5. An *integration of the text* within the cooperative process.

Separately, each component can have an impact on teaching and the decisions the teacher makes. Taken collectively you have what the authors consider an "innovative" approach to the teaching and learning of science. This study will not attempt to examine in detail all of the effects that each component has on teaching but will look specifically at those that had an effect in this classroom.

The Instructional Model

The Learning Cycle. There are five stages to the instructional model used in this curriculum. Based on a constructivist framework and adapted from the learning cycle (Atkin & Karplus, 1962), the intent is to structure learning that engages students, drawing from their existing body of experiences. Over the years the learning cycle has undergone many revisions, with the initial three stages consisting of *exploration, invention,* and *discovery* (Lawson & Renner, 1975). Calling it the "Constructivist Learning Model," Yager (1991b) describes four stages: *invitation, exploration, proposing explanations and solutions,* and *taking action.* Scharmann (1991), pulling from the earlier work of the Science Curriculum Improvement Study (SCIS, 1974), identifies three stages: *exploration, term introduction,* and *concept application.*

Each learning cycle begins with an exploratory phase, where students may be involved with reading, discussing with the teacher or peers, or engaging in an activity. The purpose is basically to engage each student's previous understanding of a particular topic and to see if she can apply her knowledge to the exploration.

In the BSCS learning cycle, *engage* and *explore* act as synonyms with the original *exploration.* Here, the first stage, *engage,* is designed to capture the students' interest and to get them to explore a topic, utilizing what they already know about the concept. The next stage, *explore*, is usually a hands-on experience in which the activity relates to what the students encountered in the first lesson. Typically, there are few instructions at this stage, thus the term "explore" describes situations in which students are prompted to explain the phenomena in their own words and descriptions. The purpose is ". . . for students to acquire a common set of experiences from which they can help one another make sense of the concept, skill, or behavior" (BSCS, 1992b, p. T26). Embedded in this stage is the need for the students to learn how to do group inquiry as a means for acquiring knowledge about science.

The next stage of the model, *explain*, mirrors the *invention* or *term introduction* phase of previous models. Originally called *concept introduction*, this phase is one in which terms and concepts are discussed and introduced. It can begin with a brainstorming session of student ideas generated by the exploration, with the teacher then helping students to narrow down their ideas or focusing them toward the intended concept or set of skills. The authors of BSCS like to emphasize that by organizing the stages in this manner, *experience precedes explanation* (a common constructivist theme), and thus, students do before the teacher explains.

The next two stages in the Five E's, *elaborate and evaluate*, are those in which students apply their understanding of the new concepts and skills to new activities and situations. Thus, these stages can collectively be referred to as *concept application*, where the application of ideas leads to understanding.

What is different with this model, compared with the others, is that the last stage, *evaluate* provides the students and the teacher with opportunities to assess understanding of the concepts and skills learned. Of course the term *evaluate* doesn't propose that learning stop here. As intended by the authors, "students will continue to construct their understanding of these broad concepts and to refine their use of the skills and behaviors throughout their lives" (BSCS, 1992b, p. T28).

Through using this instructional model, students become actively involved within the learning process. This means not only "hands-on," but also "minds-on" learning, where students are engaged in problems and discussions related to the concepts at hand. As stated by the authors (BSCS, 1992b),

> In *Science for Life and Living*, students are active learners. They question, ponder, discuss, and come to some conclusions about how their world works. They

try to make sense of things themselves and use information to build their knowledge from experience. In this way, they construct a new or refined understanding of the concepts under investigation. (p. T28)

Teaching science, or any other subject, in this manner requires a less teacher-centered approach, one that is counter to traditional practices. There tend to be fewer directions as to the outcome, and students need more time than is usual to interact with the material, to discuss their findings, and individually and as a group come to an understanding of the results of their activities. One frequent comment I hear from classroom teachers is the constraint of time. If a teacher feels inhibited by time, implementing this curriculum to its fullest may be difficult.

A less teacher-centered curriculum is an instructional plan which must match the teacher's own philosophy of what it means to learn. If she and the curricular philosophy do not match, she may revert back to her own style, and "domesticate" the curriculum. For example, one student in Donna's class actually told me that her fifth grade teacher did just that. She was told to read the book, answer the questions at the end, and turn them in for homework, with little or no discussion, clarification, or explanation. Such a "traditional" approach is at odds with the intent of the "Five E" instructional model.

Traditional science lessons also tend to focus on the facts and content of the subject and less on the overarching themes that represent the structure of scientific understanding. That is not to say that content is undesirable, but overemphasis of it tends to lead to unimaginative and frequently unrelated learning. If a teacher's philosophy is such that she believes that the content, in the form of specific terms, facts, and definitions, makeup science, then she will be at philosophical odds with the BSCS curriculum, and thus implementation will be difficult. Embedded in BSCS is the Sizer (1985) notion of "less is more" (p. 226), where fewer topics are studied, but done in more depth with an emphasis on students acquiring a broad base of conceptual understanding.

An Integration of Science, Technology, and Health. The justification for integrating science, technology, and health is to help students not only understand each of these topics separately but to discuss and learn how the three general topics have an impact on their lives. Each topic constitutes one complete unit of the curriculum, emphasized in the following conceptual manner: "science as a way of knowing, technology as a way of doing, and health as a way of behaving" (BSCS, 1992b, T19).

Chapter Five: The Science Curriculum

Science as a way of knowing is introduced first in the curriculum because it is the process of questioning and investigating that sets the stage for the remainder of the units. "The enterprise of science encourages wonder, creativity, curiosity, questioning, skepticism, perseverance, and cooperation" (p. T19). The intent is for children to learn how to learn and to learn a variety of methods that scientists use to find answers. One goal would be to alleviate the mystique of science (Lemke, 1990) and show students that they too can investigate as scientists do.

Technology, as a way of doing, is intended to show students that "people use the processes and tools of technology to solve practical problems" (BSCS, 1992b, p. T20). Each grade level concentrates on a particular feature of technology. In the sixth grade the students learn the value of constraints and trade-offs as they design the means to clean up oil from water. For example, a number of students knew that soap would clean up oil, but the trade-off is the death of almost all, if not all, of the organisms in the water. Thus, while they may use another technique that is environmentally friendly (as far as clean-up is concerned), the trade-off is that the thoroughness of the cleanup may be considerably lower.

The final feature, "health as a way of behaving," focuses more on the skills of student decision making as opposed to content knowledge. The argument is that ". . . adoption of certain behaviors and avoidance of others is most likely to lead to a long and healthful life" (p. T20). The main focus is to help students learn how to make decisions about what they eat, how and when they exercise, how to avoid use of certain substances, and to help them develop the social skills that allow them to not only resist peer pressure but also to learn how to make decisions that are good for them.

For a teacher, approaching science in this manner may be difficult. Learning science around themes is generally counter to traditional science instruction and necessitates more of an integration of scientific knowledge instead of dealing with specific disciplines, like biology, physics, or earth science. In order to integrate scientific material from several domains, teachers may believe that they need to know more about science for this approach than they did using traditional methods. However, many elementary teachers have very little training in science education and, therefore, may feel anxious about integrating science themes. Such a thematic approach may be at philosophical odds with the ideas that many teachers have who were educated in the traditions of science. For these teachers, teaching science thematically may prove frightening and most difficult.

Themes and Concepts. The final instructional component is really more of a subcomponent of the one just discussed. Just as the entire K-6 curriculum focuses on science, technology, and health, each grade level curriculum has its own theme. The major theme of the sixth grade curriculum is *Balance and Decisions*, which is then subdivided into one major concept, *Balance*, and one major skill, *Decisions*. Therefore, the focus is on *Dynamic Balance*, with students making Decisions in relation to science, technology, and health. The authors' intent was to organize the content of science ". . . under the umbrella of major concepts and skills that unify and connect different disciplines in science with the disciplines in technology and health" (p. T29). The goal was to provide students with a broad base for learning science, which of course includes some of the more traditional science content.

For example, unit two focuses on *Balance and Decisions in Science* within the science content of *Ecosystems and Resources*. Students investigate a more traditional topic of science, food chains and food webs, which looks at the resources necessary for living. In doing so they engage in activities that ask them to investigate and make decisions about what is necessary for living in an ecosystem. In the explore activity, *"Oh Deer,"* one-half of the students play the role of deer trying to attain natural resources to survive, while the other half play the role of the food, water, shelter, and/or space needed to live. As students study how populations of deer fluctuate from generation to generation, they are able to make connections between the necessity of finding a valuable resource and surviving. Through similar activities, the students are better able to make decisions about what is crucial for survival and what is not (Chapter eight will provide a detailed discussion of how this played-out in the classroom).

These three instructional pieces set general boundaries for what is taught in this curriculum. The instructional model also builds a framework for learning that necessitates a collaborative classroom structure. Thus, while the instructional and cooperative models could be considered separate, they are actually linked together to provide a framework for teaching and learning science in an innovative fashion.

Structural Components of the Curriculum

A Cooperative Learning Model. In this curriculum how the classroom is structured has a direct relationship to both the instructional and the cooperative learning model. For example, the Five E's necessitate that students collaborate, which, therefore, requires a different classroom structure than what traditional teaching and learning have to offer.

The cooperative learning model is adapted from one by Johnson, Johnson, and Johnson-Holubec (1986). In working together, students are allowed opportunities to see the points of view of others and depend more on their peers for ideas and answers and less on the teacher for direction. Inherent in a cooperative structure is the need for management mechanisms that allow for a smooth operation of the classroom. In this curriculum these skills are *move into teams quickly and quietly, speak softly, stay with your teams, take turns, and do your jobs* (BSCS, 1992b, p. T22). Training students in using these skills is an important job for the teacher, and while a less teacher-centered style may be interpreted as laissez faire, this structure necessitates a different role for the teacher, one that may be quite different from traditional practices.

Also embedded within this cooperative learning format are the use of *Team Skills*, which then become part of the "content" that students learn (hence, an overlap between instructional and structural components). For example, for unit two, the team skill is "Ask questions to help you understand one another's point of view," and the skill for unit three is "Discuss many ideas before selecting one" (p. T24). In application, the teacher moves from group to group, not only assessing student understanding of the science content but also observing the use of team skills. A strict adherence to the team skills necessitates that students rely less on the teacher and more on themselves for learning. The design is intended to ". . . help students learn to share leadership, build trust, communicate more effectively, and manage conflicts" (p. T23).

Another idea embedded within the cooperative model is that of equity. Structuring the classroom cooperatively is intended to provide all students with opportunities for engaging in activities of high quality and to promote higher-level thinking. Research supports the claim that cooperative learning has a positive effect on increased interest, involvement, and potential achievement in science for females and ethnically and culturally diverse students (see Johnson et al., 1985; Skolnik, et al., 1982; Hilliard, 1989, 1992; Lee, 1992; Shade, 1982). In general, more student interaction in a noncompetitive learning environment and cooperative learning groups allow for an increase in social collaboration. These structures then aid in the construction of shared as well as individual knowledge. Issues of science content are presented within the social and cultural context of the children's lives, allowing them to incorporate prior experiences and knowledge in the negotiation and collaborative processes embedded in cooperative learning. Through active participation, the use of nonstereotyping cooperative jobs, and the use of a text that avoids the generic use of *he, him,* or *his,* with pictures that show both males and females engaging in science tasks demonstrate to the students that all people can participate and learn science.

In using the cooperative model, the teacher and students are involved in a classroom structure quite different from a traditional one, where the teacher now allows for group process that requires a high level of student discussion. Students working together are collaborative in nature as opposed to being individually competitive, but at the same time techniques need to be builtin that hold each individual accountable for learning. Methods also need to be used that allow for smooth transitions from whole group discussion to small group tasks, and back to whole group again. The teacher now becomes actively involved in managing the class while at the same time providing students with opportunities to be more responsible for their own learning.

Active Learning. Another part of the classroom dynamics that influence how a teacher teaches is the use of hands-on/minds-on activities. The active nature of cooperative learning requires the teacher to plan for smooth transitions and operations in the classroom. This requires a lot of time and energy by the teacher. She must organize the material, where specific groups will go, how they will get materials, and a smooth transition from the group activities back to the whole group.

For planning the cooperative activities, the authors of the BSCS curriculum have helped considerably. They have built the cooperative model right into the curriculum including management techniques, the use of social skills, structuring of the group size, and activity organization. In using the curriculum the teacher does not have to decide on group size or task roles because these are already structured into the process. However, a cooperative classroom is different from a traditional one, and if the teacher's philosophical structure does not match that of a cooperative approach, the intended implementation will be a real challenge for the teacher.

Text Integration. One final component of the classroom structure is the fact that the student text is an integral feature of the curriculum. The key features of the text are, first, it is procedural where students do not need the teacher to complete activities; second, it includes all information about team tasks, team jobs, and team skills; third, it includes a variety of written forms (graphs, tables, charts, as well as poems, short stories, and expository writing); and fourth, it provides writing opportunities in which

■ Chapter Five: The Science Curriculum 61 ●

children can share their ideas with the teacher. The text encourages students to be more independent of the teacher and at the same time provides opportunities for sharing ideas with their peers. While it is possible to use the text in a more traditional manner where students read and answer the questions at the end of each activity, such an approach would undermine the success of the program.

Taken together, these aspects of the BSCS curriculum make it structurally as well as instructionally different from traditional science education. The design team, drawing from the recent literature on science education (G. Foster, Personal Communication, October 4, 1993), had at its center the concerns of both teachers and students.

Math/Science Labs

Lab Structure

The second major curricular component, the math/science labs, is a weekly, one-hour, independent investigative session for all students in the classroom. "Labs," as it is called by the teacher and the students, integrates science and math within a process of inquiry. All students are provided the same opportunities to engage in a variety of investigations that are challenging, interesting, and meaningful. Emphasis is placed on encouraging success and what students ". . . can do instead of highlighting their weaknesses" (Hart, 1994, p. 11).

By design, and echoing current literature on student assessment as well as literature on the reform of science education, numerous opportunities are provided for enabling teacher and students to assess and evaluate conceptual understanding, essential skills useful for lab completion (which include appropriate use of social and group process skills), and progress in the students' ability to explain their process of inquiry, in written and oral formats. Finally, the processes and skills essential for success in labs match the broader curricular goals of the classroom, and thus students can transfer and apply new concepts and skills attained and practiced within the lab periods to new learning experiences in and outside the school environment.

The overall design of the math/science labs originated as a means to involve students actively in relevant problem-solving activities that not only integrated science and math but also directly engaged them in an investigative process. Labs were also constructed as a means for integrating more effectively into the classroom the curriculum strands of the NCTM (1989) standards. Other central issues were those to provide students with wider choices in their learning and to provide opportunities for writing about mathematics. Donna wanted also to incorporate dialogue journals into math instruction. In a written statement she explains:

> At this point in my career, I was really interested in synthesizing a lot of things I had been learning into something [or some things?] that had a spark of originality. I was, and continue to be, really ready to experiment to put things together in a new way. I think the move to Oppenheimer at this point was crucial because I had an expectation that this was a place where I would not only not be hassled, but I would also be supported in my efforts.

Labs were based upon the following eight categories: *Patterns and Functions, Measurement, Probability and Statistics, Geometry, Problem Solving, Technology, Integrated math/science #1,* and *Integrated math/science #2,* with each lab category corresponding to a specific station in the room. At each station there are enough materials for two pairs of students. The lab instructions, including all necessary data tables, are contained in a colored folder. All folders hang from one central "Lab" bulletin board, and there are two sets of folders for each category (Thus, the need for two sets of material at each station). Labs are selected from a variety of sources, including books and professional journals, as well as from student-generated ideas and teacher-made activities that are in response to student needs. For example, if the concept of volume and the use of graduated cylinders needs addressing, Donna will either find or herself design an activity that satisfies these needs.

After each four-week cycle, the specific activities are changed, and new labs are added to the original eight topics. While students work in pairs, each student is responsible for completing six out of eight labs per cycle, with extra credit given to those who complete more than six. Two pairs of students work at each lab station and form one social group in which to discuss problems and issues regarding lab completion. When the entire group of four cannot answer a question, they "red-cup-it" and ask the teacher for assistance. Within a traditional ten-week quarter, a four-quarter school year, students have the opportunity to complete between forty-eight and sixty-four different lab activities among eight different topics. In fact, one of the strengths of labs for developing science and math understanding is that all students, even those with special needs, have a variety of experiences by which to achieve success in learning about important science topics.

All lab writing is done in a spiral-bound notebook called the "lab journal." When students get the lab folders they are required to go to the appropriate station and to sit quietly and read the lab instructions. If there are any required data tables or diagrams in the instructions, the students must hand copy these from the folders into their lab journals. Students are never given reproduced material, such as data tables or diagram templates. This practice provides them with many opportunities to make tables and reinforces the need for a table in data collection. Tables not only contain columns for recording of raw data but often contain columns for calculations of the difference between estimated and actual results, as well as ratios between two sets of data. For example, in "Going Bananas" students first estimate the mass of a banana, "mass it," and then calculate the difference between the estimate and the actual mass of the fruit. They also calculate the ratio of the mass of the pulp to the mass of the peel. When they do this with pistachio nuts, students compare the mass of the nut to the mass of the shell. The resulting ratio allows them to then consider the amount of shell that they actually pay for when they buy pistachios in the store.

After collecting the data, students are then required to "write-up" the lab. The write-up consists of the data table, any other diagrams they have created, or other data collected during the labs, as well as answers to five questions (Figure 5). The five questions remain the same throughout the year and provide the students with consistent expectations. Donna collects the journals each week, and in her written feedback to these five questions a dialogue is created between teacher and student. In her written comments she guides the students toward more detailed descriptions of the lab process and provides input as to what they did and connections that they may make between labs. Ideally the students use the teacher's feedback to improve the write-ups.

Labs also provide Donna with a wide variety of assessment (Table 2). The lab, thus, becomes more than just opportunities helping students understand math and science but is a way of helping them to become more proficient in expressing themselves better in writing. Also, because the primary language of many of her students is Spanish, lab write-ups help to promote English literacy skills.

---

1. What was I trying to find out?
2. What did I do, step by step, in order to find out?
3. What did I learn from this experience?
4. How do I feel about it?
5. What other activities or experiences does this lab remind me of?

Figure 5. Lab response questions for the lab write-up

Type of Assessment	Information Collected
Teacher Observations	Anecdotal notes to help to identify general trends in student work, help to assess the process of investigation and group interactions (attitudes, cooperation, and resourcefulness), and help to identify evidence of mastery of skills and concepts associated with the labs
Analytic Scoring Scale	Assessing the adherence to lab format. Assessing the product. Written Assessment (Within questions) Assessing growth in written communication and students' understanding of purposes and procedures of labs
Self-Assessment	Allowing students to express opinions and feelings about labs
Other	Written dialogue in lab journal, between teacher and students, provides additional opportunities to question, challenge, clarify, instruct, and reflect on lab experience

Table 2: Teacher Assessment Opportunities During Labs

Labs and the BSCS curriculum complement each other well because both emphasize problem solving within a cooperative framework. Also, unlike science fair, labs and BSCS usually have an organizing theme upon which, under the direction of the teacher, the students focus.

Cooperation and Independence in Labs

Labs create opportunities for the students to work cooperatively with each other and allow them to become more independent of the teacher. The natural progression, and the one Donna hoped for, was as follows. As a pair of students works together longer, they begin to solve problems between themselves. If they cannot, or they have questions, they then ask for help from the other pair at the station. This interaction between pairs was the reason Donna created this particular station set-up. It encourages the students ". . . to use the other partnership that was doing the lab" (Written response to fieldnotes). Asking each other questions and learning to come to consensus allowed the students to reduce their reliance on the teacher.

Donna, however, did teach her students to rely on her less and to rely on each other more. She did so in the following manner. Each station had two, plastic sixteen-ounce cups, a red one and a green one. The green cup placed over the red cup meant that all is okay, and there are no questions to ask the teacher. Red cup over green cup meant the group was stuck and needs help. At the initial stages of lab (in September) students would quickly "red-cup-it" and ask for help. Many times they did so without first asking the other pair or without formulating a question to ask. When Donna arrived at a station she would randomly ask one person, "What's the question?" If they stuttered, stumbled, or didn't respond within a few seconds, she would walk away. Doing this trained her students to follow a specific progression of problem solving. First, ask your partner; second, ask the other pair; third, formulate a question; and fourth, red-cup-it. Thus, in response to Donna's behavior, the students began to learn the system. For example, there were times when one student would red-cup-it and another would stop her and ask the group "What's the question?" As the year progressed, red cups were less frequently used and if they appeared the students generally needed more materials or wanted to show Donna what they were doing. Such a procedure enabled the students to work more cooperatively with each other and to become less dependent on the teacher.

The labs created by Donna provide a genuine view of what she believed important for learning. She structured and guided the direction by choosing the specific investigations and activities. She used input from her students, who later in the year might suggest other investigations to do. (For example, one student suggestion was used for a lab the following year.) While labs seemed openended, in reality there were answers and conclusions to many of these experiences. Labs were more teacher centered than one might think, and how she went about helping students learn provided important insight into her beliefs on teaching. She did, however, provide experiences that allowed students multiple solution paths as well as experiences for continuous practice at science and math related skills, such as the massing of objects, estimating, determining patterns and functions, constructing and using data tables, and construction graphs, and using data as a basis for making conclusions.

Labs provided the students with opportunities to be independent of the teacher in active and meaningful learning. Donna used these experiences to integrate alternative assessment mechanisms into her teaching, and she was thus able to interact individually with each student and provide positive, constructive feedback to them. For her students, labs was the most anticipated hour of the week.

Science Fair

The third, and final element of the enacted curricula, was the science fair. Mandated for each student between sixth and eighth grade, the science fair was a traditional activity in which students designed and conducted an experiment based on their interests. The students constructed a poster for the presentation and wrote a formal laboratory report. Presentation of final results was conducted during the third week of February.

Science fair is where students worked in pairs to explore, design, and conduct an experiment of their choice. With one exception, science fair had all the combined similarities of the BSCS curriculum and the labs: cooperative learning and use of social skills in a investigative process. The one component that made science fair different was the adherence to a more formal structure of a scientific investigation, including a fixed format for writing up results. Students followed a traditional written format that included:

1. A statement of purpose
2. A hypothesis
3. Materials
4. Procedures
5. Results
6. Conclusions.

As opposed to BSCS or labs, where there were general and specific topics and goals, science fair was an independent project where students were left to decide the specific direction that they wanted to take. Throughout the year, Donna emphasized that using ideas from labs and expanding on them might be a good way to find a topic worth investigating. However, getting her students to focus on the task oftentimes proved difficult. She also stressed, clearly stated in a class handout, that what they did for science fair must be an experiment. They must be able to collect and graph their data. This was somewhat different from BSCS and labs. Each of those two components, while organized around an investigative or problem-solving process, usually had an "end," meaning there was usually a concept, or in the case of a few labs, an answer to attain. Getting there, however, was one of the purposes of the curriculum. In contrast, science fair was intended to be an independent investigation, where the students are not guided to reach a specified conclusion. For the students, and Donna, it was this independent process that proved most difficult.

Science fair provided an organizational as well as an instructional challenge for Donna and her students. Up to fourth grade, science fair was done as a class project. It was only when they got to sixth grade that each student was required to participate. However, in fifth and sixth grade the students were allowed to work in pairs, thus alleviating some of the tension created by working on individual projects. For Donna this transition from group to individual projects was too quick, and it was not until seventh and eighth grade that the students participated individually and competed for the districtwide science fair.

During the academic year of this study, Donna was the science fair coordinator for the fifth and sixth-grade classrooms. Doing it for the first time not only "killed" her, but as she says, "It kicks everybody's butt." Organizing the whole event and overseeing fifteen projects in her classroom alone required that she put almost everything else, especially science, on hold. Her apprehensions can be divided into two main categories, organizational concern as well as instructional problems. Based on an interview that was specifically about science fair, Donna provided insight into her concerns as well as how these might influence her teaching.

Science Fair Organizational Concerns. By Donna's account there were a number of organizational problems with science fair. ". . . it's a random thing. You know, it isn't something that's ongoing. It isn't something that's ongoing in our curriculum and so kids do struggle with it."

Time also becomes a factor as both the students and the teachers struggled with the process.

> Part of it might be just, the, the, phenomenal amount of time it takes. And at this grade level I, I think there's [laughing] that some of the conflict is like this deep resentment I have because it, I have to essentially shut everything down and do science fair for a while. And not that that's a bad thing, but I don't feel that it gets done particularly well. And, and I keep, every year I come back to it and it's, it's overwhelming to monitor 15 projects, 15 or 16 projects. It's absolutely overwhelming. . . . It's just, you know, you can't, you can't make sure that the kids are completely on task. No matter how many deadlines and things that you, that you send out.

Science fair, more than any other curricular component, was considered imposed upon the teachers. The teachers had to do it with their kids, and because there was no math/science coordinator, there was no system in place to keep the process organized. To a certain extent, the teachers were "on their own." For Donna, that was a problem. "Every single teacher just suffers and suffers over it because it just, it takes forever, you get bored with it . . . uhm, [pause] there has to be a better way to, to set it up . . ."

Science Fair Instructional Concerns. In addition to the organization, there are instructional components that made science fair different. While a few of these features may need to be better integrated into science (for example, the notion of variables), Donna spent a considerable amount of time in direct instruction as she helped her students learn the process of a science fair type investigation. Unfortunately, this activity was, at times, not related to her other science curricula. For Donna, this lack of context created a conflict.

> I also feel, too, that we [sigh] in some ways we spoon-feed the kids and that's not good. . . . And, (pause) so, they don't really know how to do it. Maybe it's too, and it seems to me after doing this for so many years that it [the school coordinator, of which there is none] would, have a better idea about it?

A few years back, when the school did have a math/science coordinator, science fair type experiments and activities were conducted more often. The person in charge used material from the Teaching Integrated Math and Science (TIMS) program, which has at its foundation the understanding of "manipulated," "responding," and "controlled variables." Developing this understanding of variables is a weakness of the current science curricula at Oppenheimer, and thus when students began preparing for the science fair, Donna had to specifically address this in her teaching.

> I think that's something that's lacking in the curriculum, though, developing the idea of variable. And even last year, in the 5th grade, the science skill was the skill of investigation, but focused more on observation and gathering evidence but not so much in terms of identifying in quantifiable amounts, or uhm, working with variables at all. Which I think is also kind of key to the idea of investigation. [sigh] So I don't know, and for me it's just something [sigh] it's occurring to me that, that I need to, that I need to look at that and decide how to do it better. Because science fair is not getting it and it should be easier than, than it is.

Through our discussions Donna considered ways by which she could do a better job integrating the identification of variables into her current science curriculum. She realized that just as mathematics is considered a critical filter for student involvement in science, she also believed that they needed more practice with experimental techniques that reinforced the use of variables. However, she was unsure how to accomplish this.

> I'm thinking too, could I in that context [lab], instruct the students that, to identify the variables? Could I do it within the lab context? Then it would be something easier for me. Uhm, cause then it, and the important thing for me too, is that it uh, is that the students get regular practice. The reason why this is so difficult is that it happens once a year. You know, it happens over and over again but it just happens once a year and they forget. This way they would have practice once a week.

As she continued to think aloud, other choices came to mind, ones that fit better with her ideas of what was important for teaching and learning, as well as with her view of effective assessment.

> I would be just as comfortable with my kids if they simply did it like lab. Wrote it up and responded to the five questions and . . . would be comfortable with that or have, having them develop a, an investigation or develop an activity that would be appropriate for labs. And that, I think that's a fairer assessment of that and it doesn't, you know . . . you know they practiced it. Is there a way that I could do that? Could I get, you know, develop the concept of variable? Which is important, it really is.

As discussed, the one component of science fair that made it different from the other curriculum elements is that it was the only activity which was truly openended, where the students "discovered" what happened. The other curriculum components most always had answers to attain or concepts to understand. In many instances, it was the process of getting to the "end" that varied for her students, and embedded in that process was the need to practice and become proficient in social skills. For her, it was these experiences that are powerful for learning. Because of the truly independent nature of science fair it was more difficult for all students to remain interested and it was more laborious for students to develop an original investigation. These reasons alone, as well as lack of practice, may necessitate more teacher intervention, and thus, made Donna feel that she was "spoon feeding" her students to a point of overt direct instruction.

Final Thoughts

Each of the three science curriculum components required that Donna make decisions as to how she structured the learning environment. For BSCS, the structure mirrored her beliefs of what science and investigation should be. Because of this, she may have been more inclined to follow the curriculum quite closely. However, if there were instances where changes were necessary, she may have been able to do so more easily because her overall philosophy was similar to the one portrayed by the authors. If the curriculum did not parallel her beliefs, as in social studies, Donna spent a considerable amount of time redesigning the activities to fit what she believed important for her students. In doing this for social studies, the resulting curriculum closely resembled the structure and organization of *Science for Life and Living*.

Math/science labs were a representation of what Donna believed important for learning. Labs integrated cooperative learning into a wide array of investigations that provided the students with a number of different experiences. One of her goals was to have students move toward independence, to become better able to use each other for help as they solve problems. For her this is key for life-long learning.

Science fair created the most conflict for Donna. She knew that this particular process of investigation was important for her students, yet she was unsure of how to integrate it more into what she was doing in the classroom. Time, frequently mentioned by teachers, was a considerable constraint, and because of this, there seemed a greater sense of urgency to help her students complete the science fair and to prod them along to completion. I would imagine that science fair, just as in labs, was an evolving process, where over the years she would design and structure it to not only fit her comfort level but also the needs of her students.

CHAPTER SIX
The Context

The School

At the time of this study, Oppenheimer Magnet School had been operating for ten years, and as the last magnet school to be initiated by the city, Oppenheimer ". . . was opened, really at the behest of the community of this side of the expressway. We want something for ours, and we don't want it to become an elitist school" (Principal Interview). Oppenheimer not only enrolled students fluent in both Spanish and English to create a bilingual/bicultural vision but also to create an environment that had a multicultural feeling. Mr. Grouse, the school's principal, had a two-pronged vision, incorporating bilingualism with active learning.

> I would say that, probably the bottom line is that we want children to become active and empathetic learners. So that probably would be the major thrust of the vision. The other component to the vision, that we have been struggling with over the years, and are continuing to struggle with is the bilingual program. What I would say my vision for that is, is that I would want those children that are limited English proficient to graduate at or above level in both languages, and would want my child to be fluent, at least verbal proficient, in Spanish. And I think those are goals that are attainable. We're not there yet.
>
> It's very important to many parents, it's also important to them that their children be taught, what I heard often, the "right" things. And that they be able, well first off that they be able to go to high school and finish high school. I mean it's literally that basic. And also that they be able to compete, and get a good job. And they recognize thats in English. I have to give the community a lot of credit.

Mr. Grouse also talked about Oppenheimer's "magnet" status and attempts to clarify any misconceptions in associating with that term by stating "We do not hand pick our students." He continued:

> It's not the case because, what we are supposed to do in the city . . . and again, magnet schools are different in different parts of the country, we're supposed to integrate. That's all we're supposed to do. Integrate races, period. And uh, and what each magnet school does is offer particular programs that will entice a family to apply. So that while Oppenheimer is a bilingual school where we try to promote Spanish and English, and another school might promote math and science, and another school might promote the performing arts, etcetera, and then we put the applications into the computer, which works like a slot machine, it's just a random, and then it clicks and whose ever number comes out, that's who's accepted. Now the one thing that we do do, that I alluded to earlier, is we do give priority to families who already have children here.

Thus, one strength at Oppenheimer was that the school personnel made attempts to provide consistency for families. If one child was accepted, then a sibling would also be allowed to enroll. In fact, Mr. Grouse is extremely proud of the fact that families seem to like the school, and because of this, he commented that,

> . . . we're at ninety-five, ninety-six, ninety-seven percent stability. Which says an awful lot, which means chances are, [if] you start kindergarten at Oppenheimer, you're going to finish. . . . The ones that transfer out . . . tend to go to Florida, California, places like that, or the suburbs.

The multicultural perspective and the Spanish bilingual concentration related directly to the demographics of the student population. At the time of this study the total enrollment was approximately 535 students; the predominant ethnic group was Latino, mainly Puerto Ricans, consisting of 60 percent to 70 percent of the school population. The remaining population was approximately 15 percent white students, some who were ethnic Polish and Ukrainian, and 10 percent to 15 percent African Americans. In general, 50 percent of the students who enter the school at the kindergarten level were considered Limited English Proficient (LEP) (All data obtained from Principal Interview and School Demographic Analysis Sheet).

In addition to a maintenance bilingual program, Oppenheimer also had a reputation of being a cooperative learning school which was associated with the adoption of the BSCS curriculum. Mr. Grouse felt that BSCS had a positive impact on the students, providing them opportunities and experiences with problem solving in a cooperative manner. When there are student difficulties, Mr. Grouse explained, "We can sit down with them and we solve problems and talk to them. That you can't do any other place. That doesn't just happen by accident."

The Classroom

The desks are arranged in seven blocks, with five blocks having four desks each, and two blocks having five desks each. In the "front" of the room, the end farthest away from the door, there are five desks arranged in a row. These desks are parallel to the long chalkboard (The chalkboard is "L" shaped, with one long section across the front of the room, and another short section extending along the side. The two meet at the corner). This area, Donna's teaching station, is where she spends most of her day. An overhead projector is placed on one desk at the far end of the row,

■ Chapter Six: The Context   73 ●

and the images are directed toward the white wall above the chalkboard (there is no screen). On the other desks are a variety of class-related items, including student handouts, collected homework, and other material as needed for class.

Donna's actual desk is in the back corner next to the windows. At this work area she has a desk, a new Macintosh LC III computer, and a printer. Next to her desk is a small square table with small (student-size) chairs placed around the table. This is a reading and writing station and is also used as a workstation during activities in social studies, math, and science. At right angles to each other, and separating this reading/writing station from the rest of the class, are two, five-foot-high bookshelves. The bookshelves face out toward the class and contain books that constitute the class library. These books, which students may sign out, are alphabetized by author and are part of Donna's personal library. Permanently fixed along the windows are drawers, on top of which is ample shelf space for her books, teacher guides and reference material, student mailboxes, as well as supplies for activities. Each student is assigned a drawer in which she can keep her books and other classroom supplies.

The wall nearest the door has two bulletin boards. This wall is actually extended about four feet from the main wall, and the space in between the two is where the students hang their coats and keep their book-bags. On the outer wall are two bulletin boards. On these are the "math/science labs." For each category of labs there are two colored folders that contain the descriptions and directions for the activities. During lab session a designated student from each group will come up to the board and choose a lab. Lined up and attached to the short chalkboard are five cooperative learning posters. Each one describes a specific task: Manager, Checker, Tracker, Coach, and Communicator (BSCS, 1992a,1992 b). Located on the wall between two windows is another poster that lists *Team Skills* (Figure 6).

Also visible around the room are additional teacher-made posters and bulletin boards. The title of one bulletin board is "We 'Chews' to Use Our Social Skills" and is designed to represent a large gumball machine. Inside the machine are gumballs (colored construction paper cut in circles) and listed on each gumball is a social skill. In addition to the skills listed on the BSCS poster, other social skills listed here include "Use 6-inch voices," "encourage others," and "avoid put-downs." Another poster, the *Classroom Guidelines*, lists the criteria for proper classroom behavior (Figure 7). As aforementioned, practice of and adherence to the use of the social skills is one of the major themes emphasized by Donna and the

BSCS science curriculum. This theme will play a major role in the creation of this classroom's culture.

Move into your teams quickly and quietly.
Speak softly.
Stay with your teams.
Take turns.
Do your jobs.

Figure 6. Student "team skills" from BSCS

Respect the rights and property of others.
Take responsibility for your business, materials, and assignments.
Only one person speaks at a time
Keep your work area and our classroom clean.
Obey school rules.

Figure 7. Poster with guidelines for classroom behavior

Every day Donna writes two lists on the chalkboard. One is the daily agenda, and the other is the list of what needs to be done in order to prepare for the day. The agenda, which generally remains the same throughout the day, lists the sequence of events in the order and approximate time that they will occur. The other list is similar to a checklist and includes reminders of what the students need on their desk, what homework to hand in, any announcements that need to be made, and other clerical issues that need attending to. This list will be changed for the afternoon session, which begins when the students come back from lunch. Figure 8 shows an example of the daily agenda, and Figure 9 represents the list of what is needed for the afternoon science lesson.

The Students

With slight variations, the demographics of Donna's class resemble that of the school. During the academic year her enrollment began with thirty students. One student, a male, left within two weeks, and another, a female, left about two-thirds of the way into the year. A total of three students entered her classroom, all by November. Thus, her class enrollment stabilized at thirty-two students for most of the year, and dropped to

Chapter Six: The Context

thirty-one for the last three months. Of the thirty-two students, 66 percent (twenty-one) were female, and 34 percent (eleven) were male. Ethnic composition consisted of 78 percent (twenty-five) Latino, 13 percent (four) African American, 6 percent (two, one of which was ethnic Ukrainian) white, and 3 percent (one) Native American. Jokingly, Donna commented that the total number, thirty-two, bothered her because it was not a good number for cooperative groups of three, the maximum group number recommended in the BSCS curriculum.

Wednesday, September 22, 1993
1. HW Check (8:50-9:00)
2. Writing (9:00-9:38)
    /turning in final piece
3. Reading (9:38-10:26)
    /choosing a book you can read
4. Spanish/ESEA/RRW (10:29-11:14)
5. Music (11:17-12:02)
6. Lunch (12:08-12:38)
7. Labs (12:31-1:31)
8. Grading Machines (1:31-2:11)

Figure 8. Daily agenda listed on the board for Wednesday, 9/22/93

Reminder - All book bags and coats must be in the closet.
You need -
Lab Journal
Science Journal
Science book
Pen or Pencil

Figure 9. Afternoon list for Wednesday, 9/22/93

Student reading levels varied, and because of this, one of Donna's most important objectives was to make readers of all students. She even had a contest that if the class as a whole read 2,000 books, she would shave her head. While the class fell short, many students who had not previously been readers, suddenly enjoyed books. Thus, while her class was heterogeneous Donna made all attempts to create a single community of learning, one of which the students felt a part.

Factors That Influence Teaching

There are a number of factors related to the students and the classroom context that would influence Donna's teaching. The following is an introduction to a few of these factors as well as a brief discussion of how they may influence the decisions and actions of the teacher.

As previously mentioned a number of Donna's students are from a low socioeconomic background. As Oakes (1985, 1990) concludes, poor students and students of color tend to get inferior educational opportunities. Causes for this can be attributed to lower teacher expectations for these students, fewer engagements with quality activities, and reduced or nonexistent supplies and equipment necessary for learning. Also, one stereotype is that the parents of poor kids do not care about their children's education and do not help to nourish an attitude conducive to school and learning. However, at Oppenheimer, just by the application process alone, it can be argued that the parents show a vested interest in their children's education. Mr. Grouse concurred with that viewpoint and indicates that there was a strong community interest in the daily and yearly operations of the school. Nonetheless, Donna still had to deal with common urban problems, namely gangs and drugs, as well as with common societal influences, specifically the effect that living in a single parent household might have on the child's readiness for school. Regardless of these factors, Donna attempted to hold each student to her high standards and provided a variety of engaging and challenging learning experiences for all children.

Another factor that may influence her teaching is the high number of students who were categorized as Limited English Proficient (LEP). However, this impact may be more pronounced at the lower grades, where incoming students spoke mostly Spanish. By the time they reach sixth grade, most students were bilingual and thus were literate in English. (During the 1994-1995 academic year, Donna had more LEP students than in 1993–1994 [the year of this study]. Because of her use of dialogue journals for student writing in labs, Donna was able to entice these students to respond in English. Instead of the Spanish teacher's feedback, who provided very little critical support, students wanted Donna's written feedback, and thus they wrote in English as opposed to Spanish. It should be noted that Donna is bilingual in English and Russian, with Spanish as her emergent third language.)

Two other related factors that could influence Donna's teaching are the total class enrollment and the ratio between boys and girls. Thirty-two

Chapter Six: The Context

students can be considered a large class, and a large class may limit the teacher's ability to organize and manage engaging activities. In order to maintain control, the teacher may keep students confined to their seats with individual seatwork and have less student-centered cooperative type activities and investigations. Gender ratio may also affect what a teacher does. If a teacher has more boys, who may be more rambunctious than girls, then her decisions of what and how to teach may be affected. Donna did admit that one possible reason her year went smoothly could be related to the fact that she had twice as many girls as boys, and thus the boys were "held in check."

Donna is given a certain amount of autonomy in deciding how she will teach and manage her classroom. At the beginning of the year she stated that Mr. Grouse left her alone and allowed her to teach in a manner that she believed useful. Mr. Grouse explained to me that one of his roles as principal was to enable the "empowerment" of teachers and to allow them to make decisions about their specific classrooms. Thus, his confidence in Donna might have also influenced the level of autonomy she received and thus affect the decisions she made in the classroom. Mr. Grouse explained:

> She is a consummate teacher. And, uh she understands so many things, you know, not just the material or what she needs to do with the material but more importantly I think it's getting at what you were looking at and how to deliver that material to students. And uh, students that, coming from various backgrounds and uh, socioeconomic and anything that you can think of and uh, that's not an easy thing to do.

Expanding on what he means by consummate teacher, he explained specifically what he saw in Donna, and provided insight as to his views of an innovative educator.

> I think that she's uh, innovative and I think that one of the things that she does, and there's lots of things, but I think that one of the things that she does that I appreciate, that I know the kids appreciate, and I know that the parent's appreciate is that she is a sort of a facilitator. That she does involve them actively in, I mean it's not just a inane kind of, [it's] a hands-on, [it] is like whole language [it] is like all of those things that are . . . And uh, it's a real working environment and you get a sense of that immediately when you walk in there. And, uh, the kids uhm, there are many, many kids, for example, who have gotten as far as sixth grade and all of a sudden became readers because they were actually reading something instead of a ditto. So she was very good at that.

Perhaps Mr. Grouse identified in Donna qualities and characteristics that he believed important for exemplary teaching. If so, then he was confident to leave her alone, allowing her to teach in a manner comfortable for her. However, autonomy and empowerment can also have a negative effect. This can be especially true if the teacher sees the principal's view of empowerment as a way of being lazy and unconcerned with the "on-goings" of the school. Thus, the principal's respect for the teacher, as well as his management style, may have an effect on what happens in the classroom. Unfortunately, such a discussion is out of the bounds of this book.

A Special Place

Oppenheimer Magnet School may be similar to many urban elementary schools in that it has a diverse population consisting predominantly of nonwhite students from low socioeconomic backgrounds. However, unlike other urban schools (but, certainly not most or all), Oppenheimer seems to have a strong parental presence, which may relate directly to the stability of the school population. In addition, the programs offered at Oppenheimer, namely the bilingual and the integration of a cooperative learning model, may help to create positive student attitudes about themselves, others, and their school. This effect, whatever the cause, does not go unrecognized by Mr. Grouse.

> I have noticed a dramatic change in the student body over the last three or four years, and uh, and the students are the same. I mean you have the same kind of students, they're not any different than they were before, and one of the things that I'm very proud of, although it's created a lot of angst, is that we have remained true to the desegregation plan.

Thus, something is happening at Oppenheimer that the principal sees and that others may notice. Examining this potential link between the school and classroom climate may prove useful for understanding further the dynamics of a school and may be of interest for future study.

CHAPTER SEVEN
Teacher Beliefs:
A Conceptualization of Self

No matter what happens, it's all about relationships. It's all social. (Donna)

When Theodore Sizer founded the "Coalition of Essential Schools," he emphasized the fact that education is about relationships: relationships among children, their teachers, and ideas about teaching and learning (*Brown Alumni Monthly*, February, 1994). In order to understand the foundations of a teacher's practice, it becomes necessary to learn about that person, to listen to what she says is important for teaching and learning, to observe her practices, and to compare what she actually does with what she believes she should do. In this chapter, Donna discusses her beliefs about teaching in general and more specifically about science teaching. Embedded within this component of the model (see Figure 2) are her beliefs about the classroom structure, the curricula, her students, as well as the notion of her self-efficacy for teaching science. What becomes apparent is that Donna's general beliefs as a teacher and her specific beliefs on how to teach science overlap.

In teasing out the general from the specific, the discussion becomes messy. However, I attempt to paint a portrait, a conceptualization of self, of a person highly committed to her profession, one who believes in the potential of all children as she prepares them for lifelong learning. Ideally, with this portrait one component of the enactment model will begin to crystallize.

Setting the Stage: A Brief History

The High School Experience

Donna began her teaching career in 1981-1982 as a high school Russian teacher. She did this for two years but lost her position due to budget cuts and cancellation of the Russian-language program. "I was heartbroken, because I wanted to, you know, I wanted to be a teacher and it wasn't going to be . . . it just broke my heart, I couldn't even think about it." Donna was raised in a family of teachers, most notably her mother and her grandmother, and even after a stint in business, in the late 1980s, she gravitated back to the profession. At this time she reevaluated what was important to her as a teacher and she realized that in teaching high school her style was more traditional. "It was foreign language; there was a lot of repetition." She taught this way because there were so few models of how to teach differently, where traditional meant the following. (As she describes "traditional" here, Donna emphasized to me that while the "gist" of her comments is the same for when she taught high school, this description is actually in the context of her early elementary school teaching days.)[3]

> You know, using a basal, using a textbook, reading the teacher's guide, sort of planning my lesson off that . . . there might be a session of reading and then answering the questions, and that kind of thing. You know, some discussion, more teacher directed, as opposed to, [I] didn't know very much about putting kids into cooperative groups . . . . It was pretty much rote traditional, you know students writing their spelling words.

Returning to teaching, relying initially on more traditional methods of instruction, Donna decided that she needed to find out what was going on at the elementary level, and to use what she learned to reevaluate what was important for her.

A Return to Teaching: An "Undercover" Experience

When Donna returned to teaching, she purposely did so at the elementary level. Her goal was to search out many different experiences that she could use to filter and gauge what she wanted to do and how she could transfer these ideas into practice. It was these substituting experiences that gave her the opportunity to explore. Donna explains:

> . . . and that's what I sought too, as I did my undercover work, as a substitute, as I, you know, went in as an archeologist and looked around to see what was going on in classrooms.(Interview #1).

What she found was a view of teaching that was totally foreign to her.

> I was absolutely shocked. I was. I was horrified and said "Oh my god, how hard these people work." I mean I knew that my mother always worked hard, and um, I thought that I worked hard when I taught high school, [laughing] I didn't, I didn't. You know I didn't know what hard work was like until I came back to elementary. And, you know, cause I really didn't know anything about it. I wasn't happy with some of the traditional things that I was doing.

---

[3] Donna clarified this point after she had read a draft of this chapter.

However, change is difficult, and since she did not know what those changes were, she continued for a while with what was tried and true for her.

> When I first started teaching, um, teaching elementary school I really didn't know what to do so I um, I looked at the curriculum, I would um, I would do some lecture, maybe we would do reading from the textbook, there would be quizzes that would be provided from the um, um you know with the textbook publisher, I would feel some pressure to do that since that was kind of expected that those things would be in the file.

What she learned from this undercover work was that in order to teach elementary kids she would have to be highly organized and structured in a way that would attract the attention of her students. She reasoned that:

> ... high school kids, cause they're more sophisticated, if they're bored they'll let you get away with it from time to time ... because I had high school Russian students, they were pretty much, uh, the academically motivated. If they were bored to death they were polite. ... I found out right away that elementary school kids, ... if they're bored, and you don't have them, you don't have them.

Thus, there was a need for a change, and the impetus for this change came in March 1987, when she took a maternity leave position in a kindergarten classroom. It was at that time when she looked around and said: "I know nothing about teaching." At that moment she realized that you couldn't force the students into learning, because if they're bored they'll ". . . . take a mental trip, or they'll poke someone with a pencil, or they'll cause some kind of a disruption." In thinking about this, and knowing that there were unexciting times in business, she knew she had to bring something else into her teaching, something personal.

> It has to be fun for me. I have to like it ... I have to enjoy the activity myself, and I have to feel a certain level of interest in it myself. I really do. Because, I feel like I'm very enthusiastic. ... I've got a lot of enthusiasm and it catches on, and the kids get enthusiastic, they get fired up about things, they start to ask themselves questions that they wouldn't have asked before.

However, revealing the origins of this change process is less clear for her. She explains:

> I don't know if it can be pinpointed. I can start to talk about it, maybe by example. For me, how it changed for me, instead of looking at a textbook and saying

that this is the definition of what my curriculum is, I began to think about what do I think is important for kids to know, so that they can continue to learn? What kind of basis do they need? Or I'll look at a theme like, well, in terms of math I would think "well, what about fractions?" What's, and then I would, maybe I would web fractions for a while. What are all the parts of that, what do they need to be able to do, what operations do they need to know, what kinds of underlying experiences do they need to have?

It was at this time that Donna began to reevaluate what teaching was all about. In doing so, her traditional practice gave way to something else, something more personal, more engaging.

A Changing View: A Collaborative Structure

This process of change continued, and in 1987–1988, Donna was a full-time elementary teacher. During the summer of 1988 she partook in a workshop entitled "Problem Solving and Critical Thinking in Mathematics." She had heard that the program was on the "cutting edge," and she intended to be a part of it. It was this program, more than any other, that began to solidify her notion of what was important for teaching and learning.

> ... it gave me my cooperative learning background. I focused on things that um, in strategic problem solving and how to teach kids, within cooperative groups, to solve math problems and to expand their idea of what math could be, and what math, what problem solving is. That problem solving isn't necessarily word problems, "Johnny has two apples and Jose has four and they have a bunch of eggs and which are they going to throw first?"

The organizers of the workshop modeled and created a cooperative learning experience. Incorporated into the actual structure of the program were two main components of the cooperative learning model set forth by Johnson et al. (1986): *positive interdependence and individual accountability*. The workshop was also conducted longitudinally, with return sessions extending into the school year when participants met and discussed the curricula they had implemented. Thus, Donna not only learned first hand how to do this with others but also how she could transfer these ideas into the practice of her own classroom because the process for doing so was modeled for her.

However, in order to structure this learning experience into her own teaching style she knew she had to spend a considerable amount of time and energy arranging the introductory "set-up" stages. When she first tried to do this, her students' initial impression was one of enthusiasm.

Chapter Seven: Teacher Beliefs

However, they ". . . didn't have any of this experience, so it was a lot of climate setting and trust building." While Donna was cautious and deliberate as she organized the learning experiences, the structure and the theory behind the activities fit with what she believed was important for teaching and learning.

> . . . it just made, it made so much sense to me. To go from, to tie all of the components that a, the critical thinking, the cooperative learning. And it just made so much sense to me also that the social skills had to be taught. Because I would put kids together and have more "your momma" going on than anything else.

The challenge with her changing was not merely with the cooperative work centered on content but with cooperative work that focused on a process of collaboration and where collaboration became part of the content of learning.

Social Skills: The Driving Force

The teaching of social skills thus became a central force in organizing her classroom in a cooperative format. She believed that the development of social skills was crucial for her kids because it would be their ability to communicate and use these social skills that would then lead to the development of their ability to investigate and problem solve in a community of learning. Learning how to learn, and doing it socially, then became the focus of her classroom structure.

> . . . of course it makes sense to teach kids social, you know, to teach them social skills. How else would they learn them? It seems to me that what I have here, is that we have a bunch of expectations for kids to behave a certain way, but there are, if you objectify it, and you say "Okay, this is one thing that we're going to work on today. We're going to work on this one thing which says that you stay together in your team. And I'm going to give you feedback on that. Now, we're just checking one thing, which is staying together in your team. I'm going to look for that, you guys are going to practice it, we're going to talk about how it might, what that would mean, what it would involve. You'll get feedback on it." Any one kid can say "Hey, they're only looking for one thing. I can do one thing right. They taught me how to do it. When she comes by I'll practice it, and then I won't do it when she's not looking," but you know that, that's the point.
>
> You know, it made so much sense to me that we were asking kids to be totally successful without teaching them and saying "Well" [here's how you do it]. Its like the story of the flower, the woman who was a, its an Indian story, the woman who was the poor housekeeper and someone gives her a rose and she puts in a vase and she puts in the center of the table and she realizes that the table is a mess. So she cleans the table. Then she looks around and sees that the

rest of the kitchen is a mess, and that the clean table and the lovely rose don't look right so she cleans the rest of the kitchen. And so on, and so on, and so on. So just like with that one little, you know, starting with that one thing, that one thing leads to so many other things. Just by staying together we've made an agreement that we're going to work together. Just by using quiet voices you set a certain tone, in the classroom. Just by asking someone a question, or smiling and nodding, making eye contact with someone when you speak to them, it, it sets the stage, for so many other things happen as a result of just practicing that one thing. And that made whole lot of sense to me.

Donna also reasoned that providing the children opportunities to be successful, even if it was only at one "thing," was a key to increasing their involvement in learning. She reasoned that social skills, skills that her students, who are mostly nonwhite from low socioeconomic backgrounds, were the one thing they usually do not get from their urban surroundings. She argued that many of the children, coming from dangerous neighborhoods, could not go out and play and did not learn to socialize with their peers in a manner that many of us grew up with and are so accustomed to doing. Thus, social skills had to be taught in school, and for her, the choice was logical.

And that's what I will see in my kids is that they would like "Oh, okay, this is how it's done. When people get along with one another this is how its done. Here's what they do." So if you practice these skills, you can be successful at them, and get more done. And it's fun, and it's easy. And a lot of our kids too, they have no social life. They have no social life outside of, outside of school and in any structure so that they can interact socially. It is dangerous for them to play on the street. Their parents won't let them go out. So they can't have friends. You know they can't go out and maraud the neighborhood for like, like I did when I was a kid. You know like, they can't, they can't just get on their, they can't just get on their bikes the first thing and like ride all day long. We would ride all day long on our bikes with no hands. And so from like nine o'clock in the morning until six o'clock at night we'd be riding. The first one who touched their handle bars was chicken. We'd be going through busy intersections. That was it, you know, real competitive. But they can't, *they can't do that stuff, and they need to.* And it's the same thing we have here in the hall with them. With them being quiet in the hall. When do they get a chance to talk to their friends?

It therefore became clear to her that it was important for her students not only to practice how to socialize within the specific context of learning but also how to work with others within the general context of life.

■ Chapter Seven: Teacher Beliefs 85 ●

An Evolving Practice

As Donna continued to reinvent her teaching, her theory evolved to where it became more "personalized," enabling her to use it more easily. "To me it's second nature now. It's just a, it's like breathing" (Interview #3). She also began to "import" this structure into other subjects, first math, then science, and on to social studies and language arts. All the while her main goal was to provide her students with opportunities to be successful. She created an engaging environment where kids began to ask themselves ". . . I want to be successful because it's fun. How do I do it? . . . Cause now *I want to know*."

In viewing learning in this manner, Donna's role changed from one where she would provide all the ideas and answers to one where she did not need to know everything, where ". . . maybe I can make suggestions."

Reflecting back on this evolution, from high school teaching, to working in the business field, to coming back and teaching at the elementary level, Donna compared her past teaching practice with her current one, and discussed how the cooperative learning program influenced this change.

> . . . it was how I wanted to teach. It was what felt comfortable for me. And I, and I found a way to do it. As I said, I didn't know what I was doing. So I, I was floundering. I was like, I was a beginning teacher. I had an idea in my mind about, I always had good ideas, but didn't know how to get from my good ideas to help and bring kids along. And, that gave me the structure to, to help, to figure out how to do it, and the experience. Just that whole process of teaching a lesson and bringing it back, for feedback to another group of teachers, . . . helped me immeasurably . . . it did, it just, it, it transformed what my classroom was about. Coming from a, from being a substitute for a while you know one of the most important goals became, you know for me was like "I must control this group of, you know I have to rein in these wild horses." I, you know, control was important to me at that point because you know I, it is, your goals as a substitute, if you can teach anything, you're having a really great day. Ah, you know, you're having an unbelievable day, but if you can get through the day without a serious incident occurring, someone killed or injured, you're having, you know, that's a pretty good day. So my idea of what it was, of what my role was, shifted, from being totally teacher directed to yeah I make important decisions because I'm the adult but I listen to you, and that job was less about, it was not about controlling children. My job was to set up a structure so that children would be encouraged to control themselves, and to teach them how to control themselves. But, not that I had to be rigid with control. So it changed me. In, in so many ways, and my classroom is completely different. The things I do, are completely different. [Now] I hardly ever look at a textbook.

The stage was set. The foundation of where she was now, as a teacher, had been established. She taught for three more years (1988-1991) in fourth grade, before entering Oppenheimer. It was at Oppenheimer where she began to blossom. As with the housekeeper, her "table" and her "surroundings" had been cleaned and rearranged. Now was time for the rose bud to bloom.

Coming to Oppenheimer

Donna entered Oppenheimer in the fall of 1991, the year the BSCS curriculum was purchased. The books didn't arrive until November 1991, so Donna taught the first unit using the field test manual. She had no special training with the curriculum until the summer of 1992, the year of the school in-service training. However, because of her previous experience and because the cooperative learning component was built into each lesson, she was not at odds with the curriculum and felt quite comfortable teaching it. Thus, upon entering Oppenheimer, her previous training had already influenced how and what she wanted to do in the organization of science.

> ... that's how it affected me in terms of science teaching, it started there, it's like the rose on the table. It started in math and I said "Well the next, now logical place is science, and the next logical," and so when I came to this school and there was this science curriculum with cooperative learning built right in I felt, "it's okay, I'm right at home. This is, this is easy, I can do this."

This study began at the start of her third year at Oppenheimer. It was at this stage in her continuously evolving view of teaching and learning that I began to ask her what she believed important in the process of teaching, especially as it related to science. It is here where I delve into this teacher's beliefs.

Teacher Beliefs

The preceding history provides ample insight into Donna's beliefs about teaching and learning. In this section I will attempt to detail her specific beliefs about herself as a teacher as well as her beliefs about the classroom structure, the curriculum, the students, and other pieces of her teaching environment. As mentioned, this discussion can get messy, with many ideas and beliefs overlapping into other areas.

Chapter Seven: Teacher Beliefs

Choices and Expectations

> I feel like any, any moment in here is a disaster waiting to happen. It could dissolve into chaos at any moment, but, but it won't, it won't cause its too, too ... Because it's highly structured, it doesn't appear to be, but it is, it's very, very structured and I, so I spend a lot of time thinking about that. Like how am I going to structure it, how am I going to set the parameters, how am going to let the kids know this, when am I going let 'em loose, what if they don't let loose well, how do I bring, how do I round them back up into a circle? That kind of thing. So I spend a lot of time thinking about that, and that's part of my evolution how I, how I got to doing this. Is, um, it wasn't always like this. Something has happened over time, and eventually I see myself, like even more and more, in more subject areas just, kind of getting out of their way and letting them do it.

Donna's process of change was heavily influenced by her cooperative learning experience. Not only did the workshop play a significant role in how she organized the classroom, but it also affected how she organized the curriculum and learning experiences for her students. Donna realized that her traditional methods of teaching would not work with elementary students and as we will see, not only would they not work, but for her, they were not right.

> You know, if it's not, if they're not excited about it there's no way. There's no way. You've lost them and you have to change it. And you can't, like, force them, I mean to a certain point, you know, but you can't force feed them into this because they're not going to get it anyway, they're not going to do it. ... So you have to, for me, I had figure some ways to find other things to do, and then, I knew what it was like to be bored, cause I was bored in business. And I can't stand that, I can't stand to be bored. In terms of personal theory of instruction, it has to be fun for me, I have to like it, because ... yes, yes, I have to enjoy the activity myself, and I have to feel um, a certain level of interest in it myself, I really do. Because, I feel like I'm very enthusiastic, [laughing] I think, it's kind of hard to miss, you come in here and I'm hanging from the ceiling.

Donna links the structure of her classroom to what she found interesting and to what she believed important for her students to learn. Her teaching had to do with what type of experiences she decided to create. She spent an exorbitant amount of time searching, planning, and making decisions about what and how to teach so that she met the needs of her students.

> It starts more with me and I have to sort and classify all that stuff in my head, and then I begin to think "well." Then I go out and look for the search, and I look at resources, and things like that and I'll say "what, what would be a good

way to do this?" Or I'll sit up bolt, bolt upright in bed in the middle of the night and say like [snaps fingers] "Wait a minute you know I heard that thing on the news" and so I'll write that down. And, that's more, when it begins with me, those are, that's different. So, it's like I'm thinking about it and I'm using resources from a lot of different places to put together, um, to put that curriculum together. And when I'm looking for resources, I'm thinking, "Well what do my kids know, what do I feel they know, what do I need to find out if they know, what do they, um, what are they interested in, what am I interested in, what am I good at, what are my strengths, if I'm not good at something how can I get good at it if they need to know it? What kinds of activities am I aware of, where else can I look?" So I begin to pick and choose that way and I see like "Well this would flow into this and this would make a connection here," and then I start to think more of "Well, you know what, hey we're doing degrees [in math], so we could do degrees of latitude and longitude like we did with this" [social studies activity], and get them to see that the secret to happiness is an integrated curriculum.

For her it came down to the choices she made as she picked and chose what experiences she would create in the classroom. For her, this process is different from a traditional approach, and thus, the activities would also be different.

For me, kind of like how I go about it is different and so then the nature of the activities is different too. That it's not, because, certainly I use activities that other people have developed, and I couldn't possibly, you know but it's the choices that I make. The choices that I make around those activities. Um, it comes from a different place. It's not somebody telling me what to do, but it's me saying "how might I do that? How would I, how would I begin to think about that?"

While Donna did use student feedback to help her construct and guide learning, ultimately it is her responsibility to structure the learning experiences. For example, if the students had difficulty with a topic, she used feedback to rearrange what they were doing. Providing an example she commented:

"Well, I didn't teach that very well." But that's again, that's my idea that I listen to them, if it's not them it's me. I'm the adult and I know better. Or, I should. I am the teacher, and I am the one who says, who looks at things, and listens to them, and I may know more how to change it. [While] I think it's a combination [of teacher and students] . . . I think that ultimately I'm the one who's responsible, and ultimately, I am, I am in charge.

In the process of evaluating the learning experiences, she may also think about the method and how effective her approach was.

> I think it may be both [method and the kids]. I think it could be both. I think it's maybe the method I need to, to look at the method in a different way, or the activity wasn't engaging enough, or they, I made assumptions about the kids in what their background was that were erroneous.

Her students also made choices. For her the main choice, however, was whether they would "Come-on-board." It didn't matter what she did, if they were uninterested, she would not "have them." While it was her responsibility to set up learning experiences, it was ultimately the students' responsibility to decide whether or not they would participate. In explaining what "on-board" meant, Donna used an example of how Rhoda eventually made the choice to "come-on-board."

> I mean to become part of it. To participate in the experience, and to learn from it. She [Rhoda] would hang around the periphery, and look to see well, "Am I getting involved in this activity, or am, not getting involved in it? Am I going to do it, or am I not going to do it?" And I think that maybe in traditional classrooms you have more kids who say "Yeah I'll sit here, I'll sit here and I'll coast but you can't get [me] to do it. I'm not going to do it."

"On-board" meant more than just participating in a single activity it also involved engaging within the whole classroom climate. Donna explains:

> Right, within the whole classroom culture, and also within the activities that define that, that operate under that, that umbrella of what the culture is. So you [the student] can decide "Well, no I'm not going trace my fraction pieces and figure out what it is, or no I'm not going to use a grading machine to decide what my grade is," and then you know, you can make that choice. If you make that choice it's going to affect your grade, but I can't make you do it, I can encourage you to do so, I can set up a place where if you come and you try it and it's not successful you can, um, try again, it's okay. But, I was concerned that she wasn't going to get on board, and then I just see her, more and more, getting on board, uh, because she, and she's doing the things that, there doesn't seem to be any stigma for her. She will go and work on the tape recorder, she will sit and read with me, or I read two pages for, to her one page. And she, she'll do those things, because I think she finally feels that "somebody's going to teach me something, somebody sees where I am and where I want to be."

Donna expected each student to become a part of the classroom culture, "Of course, of course they're going to get on board. Of course they will. I expect, I . . . I just expect that the classroom will, um, become a certain way."

> I see kids engaged.... so that's one thing that I'll look for. Are they doing it? Are they, are they involved and are they talking about it? Do I hear them talking about it later? Do I hear them making connections? And I, I am hearing that. I'm hearing connections among the disciplines. Um, I guess I can be specific about labs. Because I feel that maybe I have a better evaluation component going in that related to science and science connected to math that I feel that, I have that, I have it worked out in my mind what I'm looking for, and how I want them to make progress. And a lot of that progress is related to expressing themselves in writing about what, what they did there, about the experience that they had. And so I am seeing that kids are talking more about, you know in their [lab] write-ups about what they're learning or more connections.

In addition to getting them on-board, Donna also had high expectations for her students. In traditional teaching, students, especially the upper-level ones, could succeed without becoming engaged. They could "coast," without exerting much effort. However, with her organizational style, it was not possible to be passive. Students had to become active, to produce, and could not fake their learning. In discussing how all students must work hard, Donna commented:

> That's ah misconception about what life is like. You have to work hard. Things that you're not good at you have to work hard at to get better at them. So, these kids [the top ones], they're accustomed to that "I could coast on this" and yet because they're not [working hard], it doesn't take a whole lot of, it doesn't take a lot of effort to turn in your grammar exercises, turn in your spelling words, or, read the science book and answer the questions. It doesn't really take a whole lot of effort to do that, but this takes more thinking. And I set a standard that I want high-quality work all the time, you can't turn in high-quality work and slack off, I can tell the difference, I know bullshit when I see it. And, then I call the kids on it, because ultimately they're responsible for it, if you turn it in this way, and I set expectations and standards for it, I expect it to look, and if you honestly don't know what to do I'll help you to do better. But I know that you know what to do.

While setting high standards and expectations is one thing, it is quite another to "train" students to meet those standards. For Donna, this not only meant that her students have a responsibility to learn but that she also had a responsibility to help them achieve this goal. She believed that both the teacher and students have roles and responsibilities in the classroom and if each does her job, then learning is the outcome. In reflecting on a student question about how many "F's" were given for first quarter marks, Donna commented:

Chapter Seven: Teacher Beliefs

> I don't give any grades, but, it's not poss-, I don't think it's really possible if someone quotes, gets an F, together we haven't done our job. . . . So, but I, in terms of expectations, I guess maybe for me I'm just, because I'm so used to, it's again, I mean my strength to size up a situation in thirty seconds. I, I feel it, I mean, you walk in and I feel it, I feel like an energy level that's, as I say visceral. . . . I really do spend a lot time at first, like "this is how you do it. I expect to see this, and you know, when you get into groups this way here's what you will do. First I would like you to do this, here's how we set it up, it might be a good idea." I might also ask the questions that, "what might be some ways that we can structure this?", and then offer those alternatives, so that they begin to think well how might I do that? Because ultimately that's what I want to see, but I also think its valuable to say, here's how I might do it.

For Donna, high expectations also coincided with providing students a variety of opportunities in which to be successful. A major focus in her classroom was to link choices, expectations, and responsibilities to a wide array of experiences that provided her students with many different occasions for success. Explaining this notion, Donna again used Rhoda as an example.

> For example, um, Rhoda got an "A" on her math test. She got a 94 percent, Rhoda in her whole life, and she got it honestly, she got it honestly, she did, she started out, she didn't know what the heck she was doing, and she just blossomed into it. She understands about angles, she can use that protractor, she likes it, it's the first A, I swear to you, that Rhoda has ever gotten in math, and I could tell. She was dumb-founded that she got an A on her math test. But it's because she had regular opportunities, and no opportunity to fail. You see that she's gonna fail, and come in some other way. You just make sure that they can't. Now, I think that had you given Rhoda a test, that looks like a test in *Math Unlimited*, she might have failed. But that isn't what we do in here, so that isn't how we assess it. Because I think that what we do in here, is more like what we do in the world, but its less like what you do in school.

Thus, providing many different experiences is one key to Donna's classroom and a key to the success of her students. In the next section, Donna explains what this all means.

Variety of Experiences

An array of learning experiences allowed for many opportunities to find out what each student was good at. Finding out what they were good at also mirrored Donna's own personal philosophy of learning. Thus, as evident here, her personal beliefs heavily influenced what she did in the classroom.

And that's a lot of my, you know, spinning my wheels, is thinking about what does it mean, how do you get good at something, how do you decide what you're good at, because that's another one of my really important goals is that, its also about kids who are at this age, because there is a huge transition they make . . . and this is the time to figure out everybody is good at something, everybody likes something different. You don't have to like math.

Providing an example, she continued:

. . . but part of it is figuring what you like, what you're good at. So having a lot of different experiences, I feel like if they have a lot of different experiences that are connected, some of them interconnected lots of them active, they'll figure out what they like and don't like. And I don't expect them to like every one like Julian today in the lab. He started out the lab like "This is stupid" and came, I wrote it down in my notes, "At 12:45 Julian says this is stupid because he didn't get his first choice." 1:05 I went over and said "What do think of this game?" "This is fun." "Julian, what happened from 12:45 to now?" "Well I guess I didn't give it a chance."

For Donna, challenging experiences promote learning for her students, especially when the experiences foster skills, helping them learn *how* to learn. Later in the year, she continued with this theme about providing experiences, and explained ". . . part of this is figuring out who you are, what you like, and what you're good at. And that's what elementary school is about. It's about exploring" (Interview #2).

Learning in Science

Content Versus Process

Linked to her beliefs about learning, Donna's views of science, especially for this grade level, were focused on experiences that engaged the students in learning how to learn. While content knowledge was important for her, the key was in knowing *how* to ask questions, to investigate, and to solve problems. For Donna, these processes are everlasting.

I feel that science content is important but I think there are lots of ways to get at it, and unless its, um, they get it when they're ready, at some point they will if they've had enough experience and so on, and [it] may dawn on them [snaps her fingers]. But I'm less worried about that, I'm less worried um, [that] these kids know what the parts of an atom are, and that kind of thing I'm less concerned about that than I am, say, are they asking the right questions? Are they wondering about the results they are getting, are they questioning if it makes sense, are they following a process, are they understanding how we investigate something, are they understanding how we experiment to find things out, are they uh, did they understand the importance of organization and good records um, all of these things? That's more important to [me] than, um, if they can name all the categories of the Linnaeus system. You know, that's more important to me, at this point. Cause they're twelve, they're eleven and twelve years old, and there's a long time to learn those things. I'm still learning them.

She continued by explaining her notion of science and while repeating earlier views, she combined and expanded on the research element of science. It is here that the link between her beliefs and enactment became quite apparent.

I think it's a way to learn things. And that's more of ah, what my conception of science is. Overall. Maybe not specific to this group of kids, but I think that it's a way to figure stuff out. That, and so there are certain processes that are related to science, that are good ways to figure it out. Like asking yourself questions, trying to figure out what you want to know. Doing tests or experiments to see if you could, you know thinking of what that would be, trying to find out what people have already learned about it. So that research element, really understanding what your problem is. Then developing and experimenting, and keeping records of what you find, and then looking at what you find, and trying to, trying to make generalizations about it. I think that's a valuable way, I mean there are correlations across the curriculum. That's why when, while I'm teaching social studies, but I'm really teaching science. It's a social science.

She continued discussing her ideas of science, and explained that her notion of science is not strictly defined by the disciplines but is centered more on asking questions and finding out why.

Yeah, that's more, it's more amorphous, it's kind of blended and [I] think that, that's, that's for me by personal preference. Um, because one of my major goals is to light fires under kids, and to get them excited about learning, wherever it is. Another goal of mine is to help kids figure out who they are, cause, if they don't figure that out, I know people, you know, [laughing] people in their thirties who haven't figured that out yet. You know it would be helpful, it would be helpful if you figure it out at this point in your life. Or at least be aware of like who you are, at this point, cause this is a hard time for kids. So one of my major goals is what are you good at and what do you like? Not what do your parents say you like. Or, what does the teacher say you like, or what does society say you have to be good at. I want you to figure out what you're good at, so I want to give you a lot of different experiences and opportunities to say "Hey, you know I like that, I could be a stand-up comedian. Hey, you know I like that, I could be a research chemist." And, you know I think that elementary school, more so than science, and as you climb that mountain, then you have to narrow, narrow, narrow more, but down here it should be, and can be everything. And that's why it's good for me cause I'm such a dilettante anyway, cause I dabble in this and dabble in that. [I] Don't do anything particularly well, other than just um, you know where that's a focus of concentration. I'm more of a global thinker, so it matches me, it matches my personality.

How she structured the learning was parallel to what she felt good about and what felt right for her. While the content of science was important, she preferred to concentrate on the process (As the BSCS curriculum does). In doing so, she integrated that process into other disciplines, such as mathematics, social studies, and language arts. Eventually, as the students moved on in their schooling, the content of science would become more important. At this grade level, she believed that learning how to learn was most useful, especially since content may change, but the process would not.

> I think that as you go on in the grade levels content becomes more and more important, and I feel that content is important, but I, whose content? What's to say, the world is changing so quickly, you know, what I teach today if I'm going totally on a content base? If I can teach you that there are ways to learn, that there are strategies to use there, or ways to think about doing something. Or, ways to start yourself, or begin, begin thinking about things. You know, read first, ask around. The structure of the labs I mean that's, how do you figure things out in life, that's how you do it, you read it, [if] you don't get it you ask somebody else, you share your ideas, that's it.

Inquiry, then, is not a hands-off process where the teacher sets up the learning experiences and the students investigate. She believed that scientific inquiry does have a structure, and it is this structure of inquiry that she helped her students to learn.

The Importance of a Structure of Inquiry

Structuring the students' learning experience is an important component of teaching, and how a teacher does this and the embedded beliefs about teaching also make up an important component in understanding teacher beliefs about learning. Donna indicated a specific belief that learning how to learn and understanding how to go about solving problems was a crucial element for lifelong learning. However, the specific issue of a structure of scientific inquiry did not arise in our discussions until later in the year. In June, as Donna and I were discussing her goals for the year-end activity, "Oil Spill," she explained why learning a structure of scientific inquiry was necessary for her children. In doing so, she first compared language arts and science and then briefly touched on one conflict she has between teaching the two. While I realize that the topic of conflicts in her teaching science will be discussed in more detail later,

Chapter Seven: Teacher Beliefs

briefly presenting this component here sets the stage for her discussion of the structure of scientific inquiry.

In teaching language arts Donna mentioned that it was easier for her to back away more and more as her students' attained control of their learning. In science, however, because she had a specified curriculum to follow, it was more difficult, and instead, what she attempted was to set up activities that provided her students with opportunities for choice and more control. In the oil spill investigation, where the students designed clean-up techniques, Donna discussed her idea *of structure*.

> It's more difficult [in science], and yet I think that there are ways that the kids do have choices and control. I have structure to the open-ended activity, but again, it can go any way. I envision that there will be groups who, ah, drain the ocean. I envision that there, you know and, taking all the water and putting it from one container into another container thinking they have essentially cleaned up the spill. Um. I'll see some of that. That'll be a question for me to ask. I think that I may have groups that get frustrated because they can't get it clean enough. I think that there will be groups that have, may try a variety of different things and will also will be very regimented. They themselves may be very structured in terms of that testing. I'm looking for that. I'm looking for them to have that kind of freedom to experiment, but knowing that experimentation also has certain structures.

It is here that she discussed her hopes for her students to learn a structure of science. To aid their understanding she explicitly modeled a process of scientific inquiry for them.

> So they've learned the structure. Right, and that's through modeling too. You know that's through the modeling that I do. [What] I mean by modeling is that, the lesson may be structured in a certain way in the hope that the children will structure their work in a similar way, or figure out that "maybe that's a good way for me to try it. Maybe I can try it that way and that will help me get out of this, you know, disorganization that I might, you know, something that's not working for me because I don't have enough, I don't know where to put my stuff. Or I don't know how to organize it." So, some of the structure in the lessons is also to help them to get them to think about how to organize things. I also think that modeling is, you know, calling attention to success, and identifying successes, and talking about those successes, and then also talking about things that didn't work as well and what we'd like to change, and ways to change it.

By modeling the structure of investigation for her students, she wanted them to begin to organize their work in this manner and thus to take more control of their learning. She also felt that in addition to a structured approach to inquiry, her students would eventually learn the importance of discipline and regimentation, and how learning was not always easy.

> It's important for them to learn that structure, it's important for them to learn that structure, and also it's important for them to learn the discipline. . . . You have to have that discipline. You have to know, like, that there's structure there. So, I think that in this particular content area what makes [it] different from, lets say language arts or like just a piece of writing, there is that kind of precision that's necessary. So I think that the structure is probably more important.

Donna continued by providing an example of why she believed it was necessary to model a structure of scientific inquiry to her students. Just setting up the experiences for the students and then letting them "explore" was not good enough. For her, the teacher has a specific role to play.

> I had this conversation with Meera [a first-year teacher] at the beginning of the year. [Meera suggested] "Well, wouldn't it be great if they'd just like, then they can explore [on] their own?" Well, they can't explore on their own, or it wouldn't occur to them. Now my kids can, now they're coming up with ideas for what they want, at the beginning of the year no, they wouldn't have been able to. They'd just be wandering around aimlessly, but now with, having some ideas that they can build on, you know they're still constructing knowledge. They're still like, putting this all together. They're the ones who still say, you know you can nudge them this way, you can nudge them that way, but they're the ones who are either going to make the connection to what we do today to what we've done all throughout the year, or they're not going to make the connection to it.

Therefore, while learning is dependent upon the relationships that Donna and her students built together, it was ultimately contingent on the students and whether they would make connections and use a structure of inquiry as part of their learning. She could set up the activities, she could model a process of inquiry, she could provide guidance and feedback, but ultimately her students must take the responsibility for learning.

> Because, as I've said before, I don't know how many of my students are going to be scientists. I know that understanding how science works is going to be important for them, and just as a way of thinking and a way of organizing things, I think that its a good way to learn. It's not strictly a good way to learn for science, but it's a good way to organize yourself for other things as well.

Donna adds a final comment about the "structure" of her classroom. Not only is there a structure to science but everything she has done has been thoroughly planned in advance. Upon entering her classroom one may see what appears to be a child-centered learning environment with her students involved in open-ended, inquiry type learning. Behind this appearance, however, is a considerable amount of teacher planning and involvement to "structure" this environment. Thus, her role as the teacher becomes more direct where she is much more than a "facilitator" of learning.

> So, is it structured, is it so structured? Yes, it's so structured. For Christ sakes it's 95, it's 110 degrees in my classroom, Bob, if it weren't structured, I'd be having a *bad day*" (Interview #5).

Importance of Science

In response to the question "Why is science important for your students?" Donna explained that within the context of an urban setting, her kids needed to be well rounded in both science and mathematics.

> It's important for these kids too, because, um, this is a public school, this is an urban school, and if they're going to have success in society at large they've got to have it. They've gotta have it. If, if, fil . . ., Oh you know all the research, the critical filters are math and science and if you don't pursue it . . . [my interruption]. You know, if you want, if you want a high paying job, if you don't want to be a teacher (laughing), um,um, . . . it's important.

She continued by explaining her need for "social justice" and how, in her opinion, a whole segment of the population was being neglected and does not get access to quality experiences in either math or science.

> Yeah why is it important to my students? It's a, well, you know, from that thing I've got that social justice thing and it pisses me off to see them, you know a whole segment of the population that is ignored and ah, and where they don't have the same opportunities and nobody seems to give a shit. That really makes me mad. But that's a cold anger, that's one I can touch, that's one where I can say "All right this anger is mine and I'm going to do something about it."

It then becomes clear that it is this process of decision making, the process of inquiry and problem-solving, the process of finding answers for themselves are all necessary for addressing issues that affect their personal lives. She explained:

> In the long run, I think it's important too because if we look at, there's gonna be a lot of information, there's just gonna be a lot of information so knowing how, using the, that scientific method to know how to learn, how to readjust, how to,

how to figure stuff out. Um, its important for them whether they become scientists or not. To make important decisions, to make important consumer decisions. To make decisions about their families, about their children's health when they have children. You know *they need to, they just need, they do need, they need to know it, and they need to, they need to be informed,* not just be informed, but be able to think and to reason and figure it out and say well "I don't agree with that that doesn't make any sense." Cause, people can use science to say all kinds of things also. Things that might not be true. . . . to manipulate and deceive to, well you can use, that's, that's just people, they're, they're capable of anything. They're capable of incredible, incredible good and incredible evil, and that's just it, you just have to accept it. But, it's important, it really is. Where, you know, what's your place in the world, how do you figure, you know, how, how do we figure this out? All of it's important.

A Perception of Her Ability to Teach Science

Considering that Donna's class is self-contained, where she teaches all subjects and the fact that like most elementary teachers she was not "trained" in science, it would be helpful to ascertain her level of comfort in teaching science. As discussed earlier, this notion of "comfort" is referred to the teacher's self-efficacy. Here Donna will discuss how she feels about teaching science, her strengths as a science teacher, and her concerns about what she may be able to do better.

Self-Efficacy

To gain an understanding into her level of self-efficacy in teaching science, I approached the issue in two ways. First, she explained to me how comfortable she felt in teaching science and described her strengths and weaknesses; second, I asked her to complete the Science Teaching Efficacy Belief Instrument (STEBI)(Riggs & Enochs, 1990), (see Appendix B). Her responses in both instances provide ample insight into her sense of self. From the interview, Donna describes how she feels about teaching science.

> I feel comfortable in it, you know, in that, in that context. . . . I think that also, at least for sixth graders, uh, I feel that I have enough scientific background. You know I've watched enough PBS [laughing] . . . I've seen enough Mr. Wizard to figure that out. And moreover because I'm a curious person anyway, and that's just how I think, and I'm basically a nerd. When a problem gets me and I'll investigate it. So, if I find myself teaching, and don't know a whole lot about chemistry, I'll take some time to figure out what I can learn about it. So, I'll

learn it right along with the kids, and if something happens that I don't know, well that's how we find out. Because that's, that's important, that's how, that's how we have to do it when you're an adult.

To substantiate her position she provided an example from social studies (which she teaches as if it were science). A day after discussing the "Tollund Man," a story of a preserved body found in a bog, Alijendra brought in a book entitled *Tales Mummies Tell* (Lauber, 87). A chapter in the book *The Bog People* discussed in detail why bodies are preserved in bogs. Donna commented:

> ... there's an entire chapter in the back of the book on "The Bog People," Tollund Man, and other things. She [Alijendra] was interested, went to the library, did the research, found out more. She came into me and the kids went "Whoa, look at this!" That was it, because they were asking me questions that I didn't know. And you know, I used that as a beginning, because I knew it was good stuff. "Oh man a dead body in a bog, wow great!" And, and so I learned more. I went ahead and read the chapter, and I learned more, so next time I know I'll be able to answer more questions, but there's probably still a whole lot more to learn. Lots of, you know, better pictures than the social studies book has.

When Donna had difficulty with teaching she would spend a lot of time evaluating what "went on." She believed that no one feature (herself, the students, the curricula, or the method) was completely responsible for student confusions. It was not always her, the children, or the method. Part of her high level of self-efficacy was in her ability to "read a situation," evaluate and reflect upon what happened in the classroom, and "move forward" to rearrange the learning. Because this is an overarching component of her philosophy, teaching science at this level did not cause her problems. If problems arise, she did the following:

> So I try to analyze what the trends are in the confusion. What they get, what they don't get. What I, what assumptions I may have made that were erroneous, where I didn't give them enough prior knowledge, or activate enough for them. ... I don't think in terms of "I screwed up" I think of it in terms of "Oh, I didn't think about that, or here's some place where I need to concentrate." Or I try to think "what can I do differently? Let's come back over, or let's come at it from a different angle. Let's leave it for now and come back to it later, after I've done this." It's more of an ongoing problem solving. It's not like "Oh, I failed at this or, they failed at this." Which I think is, either approach isn't going to work to help anybody learn, because I learn from them everyday.

Her point of view about what affects learning in science leads directly into the discussion of the STEBI, in which she provides written feedback on the questions involved. Looking at the STEBI completed by Donna (Appendix B), you will first notice that her feedback indicates that she was upset by the nature of the questionnaire. She viewed it as "transparent," "confusing," "poorly written," and "offensive." It provided no room for explanations or interjections of "maybe, but it could be this, or this could be involved." As she explained:

> These items are written as if there can only be two bipolar explanations for why students achieve in science: us, or them. Is it unreasonable to suggest that student achievement in science is not about what the students do or about what the teacher does, but rather, isn't it about what we do together? I think that student achievement is more about the relationships that we build among students, teachers, content and process. So, I can't answer these questions. And I won't. (From written response to the STEBI)

Donna's chief concern with the questionnaire was with the outcome expectancy questions, almost all she circled as problematic. "Since when is motivation separate from good teaching? . . . . How can effectiveness in teaching be separated from motivation? Effectiveness in teaching is *about* motivation."

What concerned her the most was how teachers may actually find students "at fault" for their inability to learn. It concerned her that teachers may actually think that way, and she considered these viewpoints dangerous to her students. Quite frankly, she found the questionnaire to be an insult to her professionalism and her integrity as a teacher. I believe that if she had thought that I had written it, she very well may have asked me to leave her class, thus ending this study.

Paralleling the preceding discussion and supporting the lead quote of this chapter, Donna clearly believed that learning in the classroom consists of a dual responsibility, one where the teacher and the students work together in order to create the learning climate. Yes, she is the adult, the one who has the responsibility of making decisions about the classroom structure, but the students also have choices, and thus the classroom climate is about creating relationships and ones built between the teacher and the students.

Personal Strengths in Teaching Science

Donna questioned her strengths as a science teacher and wondered if there was any difference between this and her strengths as "a teacher globally?"

Chapter Seven: Teacher Beliefs

> I think it's all part of the same package, but the things that I feel are strengths for me, when I do the science thing, when I begin to think about, think the science thing, is that I feel that I'm pretty good at reading a barometer of where kids are and what they know, within a pretty short period of time. Some things escape me, but eventually I'll see it, and I feel that I have strengths in knowing a lot, trying a lot of things that will help them to learn. That's pretty vague, but that is I have a repertoire of things that I can [do]. . . . a lot of different experiences . . . I think I can really pick and choose. I'm not stuck in a rut where I have to do things in one way. I truly believe that it can be anyway, it can be anything. So, and sometimes it just is.

She believed, as with her overall teaching, her enthusiasm and inquisitiveness promoted an exciting and creative environment for learning.

> I also think that one of my real strengths is that um, I bring a lot of enthusiasm, and I'm excited. I'm very excited to learn these things and see what the kids do with them. Ummm, I feel that I'm pretty creative. That's a strength. I think that science isn't drudgery, it's very creative, there's a process that organizes it, but it really is [creative]. Like flashes of light from time to time that say "that's how it works" and you sit bolt up, bolt upright in bed and say "Eureka, I have found it." You know, think about, wasn't the research on the double helix just somehow somebody got so far with it, two guys were scratching their and they said "Oh."

In addition to enthusiasm and creativity, Donna likes to bring out the personal nature of science. For example, in her two labs called "A la Archimedes," where the students compare the density of various fruits and vegetables, Donna likes to tell the students that when Archimedes realized his discovery he ran naked through the streets shouting "Eureka, I have found it." For her, these instances are very powerful and meaningful for learning, and thus, she attempts to portray such excitement to her students and provide many opportunities for them to shout "Eureka." This personal nature can also be viewed in her questioning, where many times she asked the students to explain their views and ideas through personal stories or narratives.

> So that's reflected in my, in my classroom, is what the, what's the personal, what's the interpersonal thing that goes on here? What's the personal side? You know. What was the life like of the person who thought of it? What did they like to do? What were they good at? . . . . So I feel that that's, that's kind of my questioning too, it's like, because that breaks it down too, and from a, you know, a kid's perspective, it's like "Hey, I have good questions too. I like, I like to ask questions, I wonder about stuff too. I could be like that person. I could come up with something, ah, I could discover something that nobody's ever seen before. I could solve a piece of the puzzle."

Finally, the discussion of her strengths circles back to one of her overarching themes of learning, that is, knowing how to "figure something out," or to solve a problem.

> I can't really approach it from a content, ah, from strictly content. Because I don't know enough content and the content I know is going to be obsolete in two years. Or if it isn't already. If it isn't, you know it may not be obsolete, but it may be, there may be just so much more to know. So it's, to me it's important that they know how to figure it out.

Concerns About Teaching Science

The first concern that came to mind about teaching science was not related to science at all but was related to time. The school's maintenance bilingual program is the central feature and thus requires a considerable amount of instructional time. Donna would like a longer school day so that she may teach more science. Currently, she alternates between science and social studies, taking about one to one-and-a-half months for each unit. She would like to teach science everyday, but because of limited time, she occasionally feels the pressure to "move on" to a new topic: "So I'll let something die before it's ready to die." This conflict is reminiscent of *Horace's Compromise* (Sizer, 1985), where the teacher is responsible for making choices, or compromises, to best fit the needs of the students, the curriculum, or even herself. (The 1994-1995 school day was extended by one hour and fifteen minutes to accommodate for the new middle school program. Thus, Donna is now able to teach both science and social studies every day. While this necessitates much more work for her, she believes her students gain because they get that many more learning experiences.)

Donna also felt that she could be better in her student evaluations. For her, assessment and evaluation are different. She can assess where the students are and provide feedback for improvement, but evaluating by assigning grades is difficult for her.

> Evaluation, to me is really nebulous and this, this is I like, I like the curriculum, but I have a hard time with it, and to say, I have a hard time objectifying and setting standards within that process, or, Okay, you've demonstrated some, that you know how to, um, identify the purpose of something, or that you can, ah, ah, it's, that's hard for me and it's an area of my teaching where I need to get better. Because otherwise, I have to evaluate, and I was so excited in seeing progress, and in my, but I don't have the progress objectified. I see kids catch fire all time.

Her main difficulty with evaluation is with the assigning of grades. While she has her standards set, there are also societal standards that must be met, meaning schools are run by grading systems of A, B, C, D, or F, and for her, these grades may be misleading.

> I have a real hard time with that and I, it's something that I've gotta just look at, for myself, but that, so that I'm clear on what my standards are, and on what, there're also, you know there're societal standards also . . . and I usually tend to fly in the face of all of those (laugh) but, ummm, that's something I wish I were better at, and I wish I had more of a handle on [it]. And my curriculum [BSCS] doesn't help me really with that at all. So that's a journey I have to travel with my colleagues. I'm not going to do it by myself.

Therefore, because she knows that with hard work she can learn the science content and then figure out ways to help her students, her weakness is not in her knowledge base. As she explains:

> I, I think it is, within my knowledge [teaching science to these sixth grade students], and, and as I'm not usually critical of my knowledge base because I always know that, I know how to learn, and I can always learn more, and if I'm not in a state of inertia, and if I'm not lazy then I can, [laughing], I can do it. But there are also things, I would work really hard. . . . [Laughs] Um, If I would, It's important that I work really, really hard at this job. I mean, I really do. I don't even realize how hard I work at it until other people say, or call me "Dedicated Donna."

Conflicts with Her Style of Teaching

Of course Donna felt conflicted within her teaching but for her, she was doing the "right thing" for the children. Teaching them how to investigate and solve problems, thus promoting inquisitiveness, all within a social nature of collaboration, were the real skills her students would need for lifelong learning. Within this process, science content will be learned.

She explained one conflict she had by telling a story involving her students from last year. She was walking in the hall, and her previous students were lined up in the hall and

> . . . someone is reading them the riot act about how they didn't know how to behave, and they didn't know this and they didn't know that, and, "Excuse me, but that was not my experience with those students."

She continued, by explaining how she just had to keep on walking.

> I didn't want to undermine what was happening . . . but I am concerned because things that I see, or things that I look for, I don't know that um, that anyone else will give them that opportunity, that opportunity to shine that way. Or so many opportunities to be successful.

Throughout her earlier comments she stated that her style was not to allow kids to fail but to provide them with a variety of opportunities to be successful. If a student was going to fail, she had "to come in some other way." She would rearrange and reorganize so the student would be successful. Her assessment is organized around positive and lengthy constructive feedback where a community of learning is built in which the students genuinely feel welcomed and appreciated. However, she often wonders if she is doing her students a "disservice." She feared that many other classes were grounded in traditional learning, a framework that not only was she uncomfortable with but one that she felt had already failed her students. She wondered if she was actually "ill-preparing" them for a schooling system that at times would "tear at" their self-esteem and focus on disjointed facts and content knowledge that may have little or no relevance to a preparation for lifelong learning. She pondered this thought.

> I ride that conflict, that, you know, "Yes, this is like, you know, this is how I do things in the world," and eventually people might, the students might be called on to do this, and, you know, even today they can do that, but I'm concerned sometimes that the tasks that they are asked to do at school are not the same, and that's kind of the conflict I have. And, that's part of the conflict. The other thing to is that, is this time, I believe, I believe this truly, that everything I do here is, much of it is an act of faith. Not everybody is going to come around now, but none of it's wasted. That, you know, someday, maybe five years from now somebody will go [snaps fingers] "When they said in sixth grade read your own copy before you turn it in." [slightly laughing] . . . and I wonder about that, because I feel that so many things come in to whittle away at their self-esteem. So many things can happen, you know, basic judgmental grading. Grading somebody after nine weeks in the term. Reducing them to A, B, C, D, F. . . . Am I doing them a disservice? Because, I feel, so this is a challenge I have, It's also a place in my teaching that I feel that I need to find better ways to help kids see that they can cross that bridge. They can cross the bridge from what they do here, to, uh, to what happens in other classrooms. Or, so, if they go into high school and the teacher just says "read the chapter, do the questions, turn it in, get a grade," if that, there are ways that they can find to make it more interesting to themselves, because for me that's part of my motivation. You asked me yesterday if it has to be fun. Well, for me it does, it really does and so, but because I'm an adult I know how to make it fun for myself and still be on task.

■ Chapter Seven: Teacher Beliefs

While she does "ride" this conflict, there are times when it seemed worth it, when she would get affirmation from a student or parent and realized that what she did helped her students for real life learning. Donna explained:

> Today a student came to me and said "my mom went to, uh, is taking the civil service exam, she's doing something to go to work at the post office. And I went with her, and there where a bunch of problems that involved discovering number sequences, like identifying a pattern." And Julietta, who was in my room last year, just sat there and did all the problems for her mom and was teaching her mom how to do it. And another man at the post office said "how are you able to do this, where did you learn this?" "Oh, my teacher taught me to do it last year, its really easy." and the guy asked for my name, what school I was from, and she gave it to him, and so on.

However, for Donna, reality of traditional schooling sets in.

> Not ten minutes later the student said "I need to come in for help in math in the morning cause I need help." So here she is in a place in life, where [snaps fingers] where "hey I can do that, this is easy, there's no problem," but in school she sees herself as someone who's not successful.

For Julietta, learning with Donna had application and meaning, and thus, she felt successful. However, in a new classroom, one that was more traditional, she was failing. For Donna, this is "the conflict." "Whereas I don't really think that what's going on necessarily outside my classroom, in the world at large, is in conflict." However, what goes on in other classes may be at odds with what she believed to be important for good teaching and learning.

In concluding this discussion regarding her conflicts with her teaching, Donna circled back to the story of her old students being yelled at in the hall. Her remarks seemed to have less to do with schooling and more to do with decency and treating others the way you yourself would prefer to be treated.

> So, that conflict, and I, think that while another part of it to for me is because they are children, it's um, they don't know how to say "Hey" themselves. "You can't do that to me." They wouldn't think, it wouldn't occur to them to do that. And um, and so that's why I think sometimes it's a disservice, because I can't stand behind every child who leaves my classroom and say, and stand behind them and say [pounding on table, voice raised with a slight chuckle] *"You don't talk to him like that, don't you do that!"*

Moving Forward:
Linking Beliefs to Practice

There are many influences upon Donna's beliefs about teaching. Mostly, though, she sees a need to teach children to learn how to learn. To ask questions. To investigate. To solve problems. To have the confidence, and the self-esteem to say "Hey, I can do this, and if I can't, I know how to figure it out. If I need to, I'll ask someone else. If they don't know, maybe we can do this together." Ultimately, it's this process, an excitement about learning, that Donna models for her students and hopes that they will carry with them to later life. In the chapter on enactment, we will view how these beliefs are transformed into practice.

Donna explained to me at the very beginning of this study that her job was to "ignite fires," to provide a spark that would kindle learning. To do this she first needs to "dry the wood." She needs to provide and then train her students to use the skills necessary for lifelong learning. In the end, she hopes that they will be engulfed in flames, ignited by a little spark of interest, and gently fanned until they may burn alone.

CHAPTER EIGHT
Enactment

A Dynamic Balance in the Classroom

Science teachers need to provide opportunities in the form of activities that engage students in the development of knowledge and understanding; and they need to provide ongoing assessment in the form of criteria, feedback, and time for reflection throughout the instructional sequence. (National Science Education Standards, Draft, November, 1994, p. 18)

The teacher's job is to evoke and sustain inquiry into fundamental interests and to show the students that funded knowledge speaks to their most basic interests. (Schubert, 1989, p. 59)

The intent of this chapter is to provide examples of enactment within the three curricular components of science: the BSCS curriculum, the math/science labs, and the science fair. In this chapter I will provide descriptive detail of how Donna uses and adapts these three components to fit the needs of her students. Specifically, I will provide portraits of lessons centered on three main themes. These include findings that support the original theme-evidence of a middle ground for teaching science that combines various degrees of a traditional and a progressive framework-as well as findings centered around two emerging themes. These two themes are:

1. An emphasis on using social skills for learning science, and

2. Teacher modeling to train students in social skill use as well as to provide examples of problem-solving strategies.

Evidence for each theme, however, is not isolated but embedded within the activity and is surrounded by examples of the other themes. Just as with teacher beliefs, this creates for potential messiness. Thus, at times I will provide my own narrative to help clarify any prospective confusions or to add more detail with hopes of making the picture clearer.

Chapter Structure

Each descriptive component is written differently. This is done for two main reasons. One, the structure of the components is different. Two, the amount of observation time is not equal across components. Because I was able to observe lessons spanning over consecutive days, the BSCS curriculum is presented as a chronology of events (lessons). In this section I use many direct quotes from Donna, all taken from the audiotapes of each classroom session where I have precise time frames of whole group to small group classroom structures and transitions. Here the writing is a narrative of the day-to-day activities in the classroom, all embedded within the BSCS science curriculum.

Labs were conducted only once per week, and while I observed the first eleven labs (eleven consecutive Wednesdays), the sessions were less teacher centered, with less whole group teacher talk (my tape recorder could not pick up individual student talk). Therefore, instead of a chronology, I grouped the data around the major theme of teaching frameworks and categorized the findings under *Guided Learning* and *Direct Teaching*. Thus, what you will read is descriptive around these two themes instead of a day-to-day chronology. Finally, because I spent less time observing during science fair sessions than all others this section is lean in both descriptive detail and categorization of events. I did, however, spend considerable time talking with Donna about science fair, and thus her personal beliefs on science fair are well documented.

The decisions that Donna made, based on her philosophical underpinnings, will ultimately structure the classroom experiences she provides for her students. The goal in this chapter is to provide insight into this structure, and to bring to light the major themes for each of the three general components of the enacted science curricula.

The BSCS Curriculum: Science for Life and Living

In review, the BSCS curriculum focuses on one major theme per year. This theme is broken down into one major concept and one major skill for each grade level. For this sixth grade curriculum, the major theme is *Balance and Decisions*, with the major concept being *Balance* and the major skill being *Decisions*. Unit one serves to introduce the year's themes but concentrates solely on the major skill. Unit two serves to introduce the concept of *Balance* within the context of science. The activities in this unit then integrate both the concept and the skill. Beginning with unit two, a new team skill is also introduced for each unit. Following is a brief description of unit one, with its emphasis on *Decisions* and a more detailed description of lessons from unit two. Combined, the two units lay the foundation of learning for the remainder of the year.

■ Chapter Eight: Enactment 109 ●

Unit One.
An Introduction to the Year's Themes: Balance and Decisions

Grading Machines. "Can you beat the clock?" It's 1:24 PM, and Donna gives her students one minute to make a quick and smooth transition from math to science. She begins to count down. "Forty seconds. Thirty seconds. Ten seconds." A couple of boys begin to help in the countdown process. "Five, four, three, two, one." When everyone seems to have made the move she responds: "All ready? Okay, pencils down. Minds on. I'll even respect that," and she also puts her pencil down.

"It's been two days, we'll have to remember what we did." Students begin to raise their hands, but Donna waits for about twenty seconds so that more "lights are on." After twenty seconds, almost every student has his or her hand up ready to begin the discussion and review of what they did two days before (This session is the very next one after the one presented at the beginning in Chapter one).

In this class, science most always begins with a whole group discussion and review of previous sessions. As she elicits their input, it is here where Donna gets at what the students already know about the subject as well as reviewing from previous sessions. Today, as was common in many of the early sessions, they review the social skills and the students' responsibilities for moving into their science groups. Donna explains: "No talking, and no stopping. It should take you two minutes to thirty seconds. Under a minute, you're doing well. You're doing great if under thirty seconds."

The move into science groups was a little noisy and took one minute and fifteen seconds. In response to how they could do this more quickly, Shaw suggests that if someone was in their way they could say "Excuse me," but Donna reminded him that they couldn't talk, and suggested that maybe they plan ahead in preparation for such problems. We'll get that down on our way back. Most of you did an excellent job. I had your eyes."

They continue, beginning where they left off from the other day, by reading pages 14–19 in the text, which describes how to make a *grading machine*. In the preceding class period each team was responsible for presenting a description of the role of one of the five cooperative learning jobs (communicator, checker, coach, manager, and tracker). During "Grading Machines," Donna informs the students that while she will be looking for the job of the tracker and the communicator, each person has to have all of the information written in his science journal. They are given twenty seconds to review their roles and fifteen minutes to complete the task. When roles are decided on, each student gets a job clip, and when the team is ready to begin, the manager goes up front to collect the necessary material needed for activity completion.

During the task, Donna moves around the classroom with a clipboard in hand, jotting down notes of what she hears and sees as the students interact. For example, she notices that a communicator from one group asked a question of a communicator from another group. Also, numerous times she is asked questions, but she responds only to those from the communicator. Upon task completion, it is this information that she shares with her students.

> Congratulations. Most groups finished within time. Let me share with you what I saw. Maritza, as the tracker, knew the exact step that everyone was on. I saw an excellent job of the communicator. Jane and Annie were both communicators and asked each other questions.

She continues by mentioning how she saw many examples of students helping each other. As a sample of this she describes how Maggie helped Jorge by saying, "Look, I'm teaching you this so you will know." When she finishes discussing "what went well," she shares her ideas of what they need to work on. She wants them to pay closer attention to staying with the groups. That means both physically, and mentally, and she doesn't want their eyes to wander from their team and the task at hand.

Now that everyone has made a grading machine and has practiced with it in their team, Donna asks them to use their machine to make a report card. She gives three minutes to complete the task, and at 2:02 PM, she raises her hand and begins to give instructions about finishing up and going back to home-base. Once they are ready, she challenges the class to make a good transition. "Let's break the record" of moving quickly and quietly she states, and as she drops her hand they scurry from their science stations back to their original seat. In less than thirty seconds they are seated and quiet. With a thumbs-up signal, Donna exclaims "You did it!" The class erupts in a loud cheer. She then goes on to mention that she hopes the judge doesn't close school tomorrow, *"because we got some learning going on!"* The students respond with louder yells and cheers than before (Because of a contract dispute, this urban school system began classes one week late and almost had two other shutdowns).

It's 2:10 PM, the end of the day, and the students collect their belongings and begin to line up to go to the school bus. Donna walks out the door, and then back in, turns to her student teacher, and gives him a high five as it to say "Job well done. We did it," She then comes over to me and mentions: "They've [the students] done this before."

Introduction to General Patterns of the Class and the Remainder of Unit One

"Grading Machines" is lesson number one of unit one. The intent of this unit is for the students to practice the general cooperative learning skills *(moving into teams quickly and quietly, speaking softly, staying with teams, taking turns, and doing your job)* as well as the general social skills within the context of decision making. There is no science content in this unit. The purpose of "Grading Machines" is to help the students understand that a decision making process of assigning grades is better than using a machine that gives random assignments. The lesson also provides students the opportunity to practice their team skills as outlined earlier. The theme of Decisions is central to this unit, and subsequent lessons are designed for the students to continue practicing decision making in a group process. The lessons also help to create a foundation for use of the social skills throughout the remainder of the curriculum. From this lesson you can see that Donna also places a special emphasis on transitions (which she times) as well as on individual team skills. As evident in her beliefs she deems that, in their entirety, these skills are imperative for effective decision making and problem solving and are applicable to everyday life. Thus, she spends a considerable amount of time training her students in their effective use.

In this lesson, and ensuing ones, you will notice a number of patterns. For example, in a whole group structure she tells her students what she will look for. When they are in small groups she walks around and writes down notes of what she saw, and when they get back in a whole group structure, she shares with the class what went well and what they need to work on. Donna also trains her students to comply with a quiet signal, which is her hand raised in the air. This indicates to the students that they too should raise their hand, be quiet, turn and face her, and wait for directions. To hold them accountable she also times their compliance to this signal. One final pattern to identify now is her adherence to short time blocks. When sessions move from a whole group to small groups, or even pairs, Donna *always* provides a time limit. To keep the lessons on a brisk pace and to hold her students accountable, many time limits are often short, from one to two minutes. Again, she times these, and in most instances she will pay close attention to this schedule.

Following is an example of social skill training from lesson two. Again, this lesson centered on decision making. Before they begin Donna tells them what she will be looking for.

> [I will] Be looking especially for the social skill of staying in your team. Staying in your team means that you're all together and that you're not talking to somebody over here, looking around, or getting up and scooting around. All right, we're going to be practicing quickly and quietly today. Does everybody know where they need to report?

To check if they understand what their responsibilities are during this activity, she asks, "What are we especially working on today?" One student responds: "Silence" and another mentions "Staying with your team."

After the lesson Donna also provides feedback of the successes she saw as they related to staying with the team. In doing so, she provides examples of what went well and what they could work on to improve their social interactions.

> I said I would be looking for specific examples of how you stayed together in your team. I noticed in this team [pointing to one group], one technique that they used to stay together was that they asked each other a lot of questions. Sugie, I heard you asking well you know "what about this?," and taking time to explain. There was a lot of eye contact with each other so I could tell they all were working on the same thing. They spent time also reading back and forth to each other, and you could see that they were really working as a team. Now in other places I didn't see as much teamwork. What I saw was three people sitting together and doing an assignment, but they were all working separately as opposed to working as a team. You're going to be more successful [with] the more ideas you get so you're going to need those teammates.

As with the example of Sugie, Donna singles out specific instances to provide specific positive feedback of what went well during the class activity. She uses these examples as models of what she considers exemplary use of the social skills.

> Over here I saw some wonderful things going on, and I just want to say that I'm going to put Annie on the spot because, in order to keep her team together, she asks a direct question. She was asking questions of the other team members. She also took it on herself to explain, and she explained something, when it didn't get through in English, she explained it in Spanish, and she explained it again in English, and she went from one language to the other, because it was important to her that nobody in her team got left out. And Omar, when you understood I saw you get a real big smile on your face. And you made contact. So, three cheers for you that was real teamwork, you didn't let anybody get behind.

In her feedback of what needs to be worked on, Donna does not mention specific names of students but just the general behavior. "Now I also did see a number of people looking over and making comments to other people in other teams. That's an example of not staying together." In structuring her feedback this way, she models the technique of positive, constructive criticism and emphasizes the good that happened and what can be done to achieve success as opposed to dwelling on the negative and what went wrong.

The final component of the day's lesson, moving back to home base, was completed in less than thirty seconds. The students responded with muffled exclamations of "yes" and "all right." Donna tells them that during the transition she saw one student tell another to "shhhh." In responding to this example of being quiet, she comments: "Keep'em quiet, keep your teammates honest. That was great."

The preceding examples are all embedded within the enactment of the BSCS science curriculum. "Science," in terms of content or a process of investigation, has little emphasis here, and thus plays no role during these activities. However, for Donna, and obviously the authors of the science curriculum, building the foundation of good social process, being able to work together in an organized and calm environment, and learning how to interact during a decision making task are crucial in the students' ability to address science and science related issues. While similar to the written curriculum, how Donna structures these experiences is highly influenced by her personal philosophy of what is imperative for learning. Following will be descriptions of lessons from unit two, where more traditional science content is evident as the students continue to practice the skills introduced in unit one.

Unit Two.
Ecosystems and Resources: Balance and Decisions in Science

Unit two is where the overall theme of this sixth grade curriculum, *Balance and Decisions*, is first used together. This unit is also where the lessons initially focus around scientific concepts. In this unit the concepts are *Ecosystems and Resources*, and the students apply their decision making and social skill prowess to topics that relate directly to how organisms interact and survive within their environment. Therefore, while there is still an emphasis on working together and making decisions, the context of this work will for the first time involve scientific content knowledge. However, as a reminder, the focus of this study is on

Donna and how she structured the learning and not specifically on the students as they learn science content.

This section begins on day one of unit two. Because of the school's daily structure and the limited time block in the afternoon, when science and social studies are most often taught, Donna alternates between these two subjects, taking about one to one-and-half months per subject. The students have just finished a social studies unit and are now going back to science. Because of this lapse of time Donna and her students take about ten minutes to review where they've been in science.

An Introduction to a New Social Skill. It's 1:31 PM. Donna: "Okay, thirty seconds please. Get out your science journal and your science book. A pencil and nothing else on your desk. Pencil, science journal, science book, thirty seconds, beginning now." Occasionally these transitions take longer than expected, and sometimes the extended time is unacceptable, with Donna informing the class that they are taking too long. However, in other instances, like today, she allows for the extra time so that all materials from the previous lesson may be put away. Thus, thirty seconds here turned into four minutes. Once they are ready, Donna begins.

> It's been a while since you've looked in the science journal. Maybe it's not so fresh in our minds, what we've done. I'd like you to take about two minutes right now, and not go through the book, but I'd like you to look through your journal. I want you to re-read [the entries] what you wrote, up until the end of the journal. The different activities that you did, I want you read through the journal, take two minutes. At the end of the two minutes be prepared to share what you remember with a partner.

(The journal is where all writing goes, including book and teacher generated questions. Reviewing the journal will, therefore, act as a map of where they have been.)

After this two-minute session, Donna gives the class one minute to review with a partner what they remembered. It is only after this session that they take time for a whole class discussion of what they did in unit one. During the whole group session, individual students provide their input of what they remember from the first unit. Jeanine begins by saying that she remembers "Grading Machines." She then goes on to explain how they used the machine to randomly assign grades. Donna comments, "So you learned something about the fairness and certain decisions shouldn't be made in a random way?" As they continue, Jessica mentions the IDEA method and explains, "That it helped you, like, um, understanding, like, problems or something" (IDEA stands for: Identify the

■ Chapter Eight: Enactment   115 ●

problems, Describe all the options, Evaluate all the options, Arrive at a decision, BSCS, 1992b, p. 3). Donna asks for more clarification and an example of how they used the IDEA method. In doing so she orchestrates the group discussion, working her students to provide more detail and to link their ideas to what others say. Jessica explains a little about one of the problems in the book, and again Donna paraphrases, "There was a problem listed in the book that you could use the IDEA method to solve them. So, how did that help you to solve that, that particular method?" Jessica said that it simply showed you how to do the right thing. Joko mentions the problem that focused on deciding what movie to watch. Donna asks him to relate that to what Jessica said, and he responds that it was a problem and the IDEA method was used to solve it. Another student comments about the very first activity she did, writing down what she thought science was (this activity was not part of the BSCS curriculum). Donna responds, "That's interesting. How do you feel about science now compared to what you wrote about the first day?" Jorge responds, "It's better." Donna: "What do you mean it's better?" He goes on to explain that they don't just do it out of a book, but they "make things up" and write it down in paragraphs to describe what you're learning. Donna: "So you feel that you're writing and learning more as a result [of this experience]?" She then mentions that it may be worthwhile throughout the year to go through the journal to see how "our minds have changed. We used to think something and now we don't think that any more."

In continuing with this review of unit one Donna asks the students how all of their examples are the same. She wants to know what they all have in common, about deciding which movie to see, trying to decide what to do about the big homework assignment, whether to go to the soccer game or to a friend's house, and how the IDEA method fits in here. The class gets real quiet and seems unsure of what to say. With a pan balance in her hand, Donna moves from behind her work area, walks out in front, and sets the balance down on a desk. "What does all this have to do with the science theme this year? That's about balancing and decisions?" As she speaks she moves the pans up and down and then uses her hands, in an up and down motion, to resemble an idea of balancing. Jeanine responds that they were asked to pick a problem to solve, and they can take all the options and balance them for "how many good ones and how many bad ones." Donna responds:

> Okay, so, let's say you pick out six options using the IDEA method. You've decided on six different possible things you can do? You can balance them and say "well, I like this about this one, but I don't like this about it." So the scale moves up and down depending on what the pros and cons, or [what] the cost and benefits are of using those methods.

She then briefly comments, with excitement, on how much they remember, and without stopping, she begins to introduce the main goals for unit two. "We're going into a unit that also focuses about balance, and about decisions, but in terms of science and talking about how different things interact in an ecosystem."

During this introduction, the class, with Donna's guidance, spends about ten minutes reviewing what they did before in science and how they may create a link between the themes of balance and decisions, to science. Helping them make connections is an important goal for Donna. While the discussion seems formal, with the students being asked to recall previously learned material, the session is rather calm and frank as the students speak freely about what they remember from earlier experiences. It is during these discussions that Donna hopes that her students will make connections between what they are doing now and previous experiences. It also provides her an opportunity to gauge what the students know about a particular topic, and what they remember from earlier lessons.

After this review and discussion Donna asks the students to read page 47 in their textbook, which is the first lesson of unit two and which sets the stage for the introduction of the new team skill. When they are done Donna introduces what the unit is about.

> The focus of this unit is specifically [pause] about science. Okay? The other units that we'll be working on, will also integrate the ideas of technology and health, but this one focuses specifically on science, and scientific method.

Before they begin to discuss unit two, Donna again diverges from the textbook, and she asks what they learned from last year. Most of the responses center on the idea of gadgets made and facts learned, and not one student mentioned the process of science. To probe on this idea Donna continues.

> We use a scientific method to investigate something. Now when we use a scientific method to investigate something, how do we get evidence to prove what we think we see? How do we get evidence? What does it say in the book? How do we get it? What are some ways that we can find out about something?

■ Chapter Eight: Enactment 117 ●

Joi says that observation is a way to get evidence and mentions that watching and listening are ways to observe. Donna creates a list on the board (See Figure 10) and expands on those ideas mentioned by Joi.

```
Observation
watch    touch
listen   ("taste")        [Donna recommends against tasting
experiment                because it's not safe]
```

Figure 10. Student-generated list of ways to observe

The class is real quiet, and Donna continues by stating that there are additional methods, other than experimentation, that they can use to find something out. Donna: "Look at the work of other scientists, read and observe what they've done, to see if we can learn anything from them also." She then asks for other ways to observe, with "taste" and "touch" being provided. After a short pause, she asks them to open the text and read pages 48 - 49, which introduces a new team skill, which is "Ask questions to help you understand one another's point of view" (BSCS, 1992b, p. 48). During this unit, Donna will look for evidence of the use of the new team skill, and in preparing the students for this new skill Donna provides feedback on their use of the general team skills.

> You've gotten so skilled at the basic team skills, you're moving quietly and using six-inch voices, staying together as a team. We're going to upgrade, and teach another skill that will be beneficial for you to continue to work together in groups.

Before they read this section though, Donna continues to draw their attention to the notion of observation and why it may be important for scientists to work together as a team. Without stopping she asks the question again, paraphrasing so it's different from before. Even after she asks a specific student for a response, she again continues to ask the question once more before allowing the student to answer. Donna asks:

> Lueere, what do you think about this? If you're trying to figure something out, and you're using observation as a method to do that, like you're trying to [take information with your senses] why might working with a team be helpful?

Lueere responds that there would be more opinions and ideas with more people, and Donna suggests that they also have an extra set of eyes. If something gets missed they have an extra person who may not miss an observation. Donna responds, "So if it's going to benefit our learning in science, to work as a team, we should learn how to work as a team better." They now read the pages to review the new social skill ("Team skill" and "social skill" are terms that are interchanged by Donna. Specifically, though, the social skills are general, and team skills are introduced in each unit, beginning with unit two). When everyone is done reading, each student takes thirty seconds to explain the new skill to her or his partner.

The pace picks up here and the students stay with her. She now gets them back to a whole group session and asks for a couple of volunteers to read about the new social skill. Many students raise their hand, hoping to be called on. Maggie is chosen and explains that by using this skill, it gets everybody in the group involved, which is good because everybody has ideas. Shaw comments that using the skill is a way of not having to rely on the teacher; you can ask questions of others. Jennet mentions that when the group gets stuck, another person's point of view may help solve the problem. After these responses, Donna continues.

> You've probably seen it used a lot, especially by me as a teacher. . . . Sometimes when you give me an answer, what do I say to you? [One student responds "No." The class laughs.] I might ask you to ask your teammate. What might I ask you?

Student ideas include "Why do you think that? What do you think about that?" Donna adds her ideas. "I might ask you 'can you tell me more about that? Can you explain it to me what you mean?'" She then uses a specific example to model how she used this skill to ask someone for clarification.

> I'm thinking of one specific example, way, way, way back at the beginning of the year when shapes, about geometry, and one of the groups said, Jorge I think you were in this group and I think Alijendra you were in this group, and you said geometry was like science. And when you first said that, I didn't understand really what you meant. So, "what are they thinking about here?" Cause all of you, your heads were nodding, you all believed it. And I think I asked you "so, how is it like science?" You guys explained it to me and it helped me to understand. You guys remember that? I remember it very well.

With only a few comments from her students, Donna then begins to describe that during this unit they will be working on the scientific method, observation, and practicing the new skill. (She is moving quickly, from one topic to the next, and the class is keeping up with her.) She

Chapter Eight: Enactment 119

emphasizes her responsibility during class sessions by explaining how she'll walk around, with the clipboard in hand, and look for evidence of the new skill.

To help her students recognize the new team skill and before they continue with the unit, they do a short activity called "Looks Like/Sounds Like." Donna asks for specific ideas of the new skill that will fit within the two categories. On the front board is a large (2.5 feet by 3.5 feet) sheet of writing paper with the two categories already labeled. To focus the class into generating ideas to include on the poster, Donna poses a hypothetical situation.

> If I were to look for evidence, Julian [getting his attention], in you practicing the new social skill of asking questions. Let's say I'm not in the classroom, but I'm standing in the doorway, what kinds of things might I see that would let me know as clues, please put that away right now Joko, that let me know as clues that this was what was going on?

As students' ideas come forth, Donna writes these on the paper (see Figure 11). After "Looks Like" she asks the class to tell her what she would hear if she weren't in the room, but was listening over the classroom intercom. One student suggests, "Ask questions." Donna probes, "Can somebody get specific about the kinds of questions I might hear to help me understand what somebody else thinks?" They continue, and even though the session has lasted 25 minutes, the atmosphere is relaxed and quiet, with Donna and her students occasionally laughing as they generate ideas. When completed, Donna reminds the class:

> As I go around, and I look, and I start to see whether or not we're practicing this skill, these are the criteria that I'll use to see how you're doing in developing the social skill. These are the things that I'll be looking for.

It is now 2:05 PM; class ends at 2:10 PM, but Donna continues by introducing the job of a Coach, one more team component that will be used throughout unit two. She asks her students to go back to page 9 and to "re-read the description of what the coach does." She gives them thirty-five seconds and then asks for a volunteer who can explain the job of the coach. As a class they then generate ideas of characteristics of a good coach. Donna writes these ideas on the board, directly underneath the Coach poster. Student-generated characteristics include keeping people on job/task, practicing the skills themselves, and not being bossy. Donna comments, "The best coach is one who does things by example." She then asks, "How many of you really like to be bossed around? Raise your hand if you love to be bossed around. [Zero hands.] I can't stand it myself." The class then digresses into a conversation of bossiness. Many people are talking, including Donna, until finally she raises her hand to bring them back together. It's 2:09 PM, and a few students keep responding about the bossiness issue. One male provides an example, "Don't do this, don't do that," which generates laughter from the class. Donna: "You know what? This year we'll be teaching you some skills that will help you also to deal with people who are bossier." One student comments, "How 'bout my mom?" and again the class digresses into numerous conversations. Donna, talking over a chorus of conversations, "On that note, we've identified this. We'll come back to this tomorrow." Amidst the chatter, she tells them that "today we've done a good job. Close it up," and get ready to go home.

Ask questions to help you understand someone else's point of view.

Looks Like	**Sounds Like**
Showing	"Now I get it"
Pointing	Questions
Faces excited	"Do you get it?"
Nod head	"What do you mean by. . .?"
Eye-contact	"Can you explain it to me again?"
Actively listening	"Do you think that's a good idea?"
Heads together	
Gather over work	

Figure 11. Class list of ideas for "Looks Like/Sounds Like" of the new social skill. (The list remains posted on the classroom wall for the entire year so students may refer to it when they wish to look for reminders of what this particular social skill may look like, and sound like.)

Practicing Social Skills in the Context of Science: A Day in the Life. "A Day in the Life" is the first activity in unit two and emphasizes the "interactions among the living and nonliving components of three ecosystems" (BSCS, 1992b, p. 75). The three ecosystems, a salt marsh, an oak forest, and a rain forest, are presented in picture format in the textbook. Each team member reviews one picture story of an ecosystem and makes a list of all living and nonliving "things" in the pictures. From this list each team member is to then identify and list examples from each of the following three sets of interactions: living to living, living to nonliving, and nonliving to nonliving. When completed, and while still in their teams, each team member describes to the other two members the interactions in their ecosystem. During these short presentations, the students are expected to practice the new social skill; "Ask questions to help you understand one another's point of view." The team member with the new team job, the coach, is responsible for documenting "how many times the team uses the new team skill" (BSCS, 1992, p. 77).

"A Day in the Life" took two days to complete. Presented here is a brief description of day one followed by a detailed account of day two. The first day, Donna spent most of the period helping her students better understand the ideas of "living" and "nonliving" components of a system. They began by generating ideas about their classroom system and then linking their ideas to three general categories of interactions: living-living, living-nonliving, and nonliving-nonliving. Initial problems arose as students were confused with the ideas of living and nonliving, but with practice they were better able to generate examples. The entire period, 1:33 PM to 2:11 PM (38 minutes), was spent on this topic, and all of the student work was done at their seats, either in a whole group (teacher directed) or small group structure. There were problems of students talking, of not paying attention, complaining about the composition of their science teams, and not moving to teams quickly and quietly.

At the end of day one, Rebecca quietly complained, "We didn't even do science," and Donna admitted that her kids were not "on it today." She was disappointed that they were unable to get to the actual activity and made plans to move through the activity tomorrow. One of the reasons for Rebecca's disappointment could have been related to the fact that they had just spent the last two days in their seats instead of being able to spend more time in their science groups. Usually, they get the opportunity to move into teams and work collaboratively on the lessons for an extended period of time. Today, they moved into teams for less than ten minutes before they had to move back and get ready to go home. Also, the students had just finished social studies and, because they were having so much fun, didn't want to go back into science. Donna mentioned to me that this happens frequently, where the excitement level is so high at the end of a unit (science or social studies) that the students want to continue. She explains:

> You know it's funny, the kids ask "well when are we going back to science," and I say "Well we'll go back after, after, um, after Thanksgiving." And do you remember when I said "well we're going to leave science now and go to social studies" and the kids said "Ooooooooooo" and now we said we're going to leave

social studies and go back to science and the kids went "Ooooooooo" [slight laugh]. (Interview #3)

Even still, Donna's comments after this day indicate that she needs to get them more active tomorrow.

On day two of "A Day in the Life," the supervisor of Donna's student teacher was in to observe. Thus, there were four adults in this classroom of thirty students. This doesn't seem to bother the students. Donna's class has a "revolving door" with many people coming in to observe. Donna comments, "The kids are used to it."

The session begins at 1:27 PM with Donna asking for feedback on problems with yesterday's transition from labs to science. Shaw mentions that there should be no talking, and Joko suggests that they need to be quieter with their chairs. Donna asks Joko, "Okay, so, how can we solve that problem?" One suggestion is to wait until everyone settles down and then move the chairs. Another is to leave all chairs at home base, and sit at available chairs at new seats. Donna asks, "How many of you think that that will work better if you just sit at the chair [that is at the station]? All right, I think that might be an improvement." She then reviews a problem with a specific station, one where there is only two chairs, and she suggests that the group that goes here use an empty chair from the station that is next to where they are sitting. After reviewing one more problem Donna comments:

> Since this a social skill that we've been practicing for some time, we should be mastering this, and thinking of ways we can continue to get better at it. All right? Let's practice now quickly and quietly [students eagerly begin to get up and move]. Excuse me, just one minute, I don't want you to get up yet. Please wait until I say go. Okay? I'm still giving instructions.

Before they move she reviews where the ten different stations are, and indicates to one student who was absent, what station she should report to.

After the move she asks the class, "How do you feel about that?" After the students self-assess the move she continues with her feedback. "That was a marked improvement [it took forty seconds]. What was better about it?" Rhoda says it was quieter and they moved quickly to the team location. Maggie says that they knew exactly where to go. Rebecca states that they decided, beforehand, whether or not there would be a problem with chairs. Jennet comments that the movement was crisper;

people weren't bumping into each other as they moved. Donna even mentions that she saw one student use the job as the coach to remind another student, who was talking, to stop so they could complete the task of moving quickly and quietly. When finished with student comments Donna tells them, "Once again you have shown me that you do have this skill mastered. Which is, umm, which is very exciting."

At 1:34 PM they begin with each group reviewing what they did the day before. After thirty seconds for review, Donna begins the discussion by asking what they will be doing for the day. As students respond, she writes the procedure on the board. The sequence is (a) personal "field trip" (the review of the ecosystem picture story), (b) tell about your "trip," and (c) lists of interactions. Within each category, Donna asks for more clarification of what they need to be looking for. It is here where the students mention that they need to be looking for evidence of the three categories of interactions, they need to tell their teammates about their trip, and then, as a group, they need to make a list of these interactions. Within the group process, Donna allows team members to choose which jobs each will do (communicator, tracker, or coach). When the roles are decided on, she hands out the appropriate job clips. She then gives them ten seconds to review and to tell their partners which section they'll be in, which ecosystem they will become "experts" in. Next, she gives them three minutes to quietly, and individually, review the picture stories of each ecosystem. When the three minutes are up, "I will call time for you to take turns explaining [about] the interactions in your particular system. Three minutes beginning now!" Almost immediately after they begin, she reminds them that each individual should be reviewing quietly; thus, there should be no talking. After one minute, Donna intervenes and states, "Attention! Attention! I'm going to give these instructions again. You now have two minutes to look over your things silently! You should not be talking."

After the initial review Donna gives them four minutes per person to explain what they saw in their picture story. She wants them to concentrate specifically on the interactions that they discussed. Donna explains:

> Remember, that as I walk around, I'm going to be looking for examples of you asking questions to help you understand someone else's point of view. Your coach in your team is also going to be taking, ah, keeping a tally mark for every time that they hear someone in your group ask someone else a question about something that they said. So for example, if Omar says something and Lueere doesn't understand what he means and she says "Well, can you explain it to me again?" If Elizabeth were the coach she would make a little tally mark here in

the tally area. At the end of the twelve minutes of discussion, we will, you should total up how many times you've practiced the social skill. As I go around with my clipboard, I'll also be listening for specific examples, like the ones you that you listed here [pointing to the "Looks Like/Sounds Like" table] that will show me that you're beginning to practice this social skill. Before we begin, you should decide who's going to go first, who's going to go second, who's going to go third. [a six-second pause] Have you all decided? [Students respond with a choral "yes"] All right then, you may begin [it is 1:47 PM].

As they begin she hands out the social skill tally sheet (see Figure 12) that the coach will use to keep track of the new social skill. She also briefly stops them to suggest, as Maggie had asked, that when they explain their trip they should use their book to show their team members the pictures and the interactions they came up with.

After two minutes, Donna raises her hand for quiet and asks how many groups have moved onto the next description. She suggests that those who haven't done so should finish up and move on. She then announces that they have another four minutes to explain the next one. Cutting in early like this is one way for Donna to keep the students on track. Keeping them on a time schedule requires that they stay on task and "stay" with their team.

The students are enthusiastically engaged during this small group session. Heads are huddled together, listening and looking at the pictures. One student usually talks at a time. The noise level is high with students laughing, and flipping pages in their books to show and explain their fieldtrip to their fellow teammates. Donna walks around, clipboard in-hand, hovering over groups to listen and write down notes of what she hears and sees.

At 1:52 PM she raises her hand for silence and checks to see where they are in this process. Finding out that they are moving along quickly, she tells the class that they have three minutes to complete the third field-trip. At 1:55 PM she raises her hand again and asks if any group needs thirty more seconds. Those who are done should go through it again. At 1:58 PM, thirty seconds after she raised her hand to get their attention, the class got quiet. Now back in a whole group session, Donna asks Jeanine, her group's tracker, "What step would we be on in the instructions at this point?" Jeanine is unsure, and Donna poses the question to the class, "Where could you look to find out?" Students begin flipping through the book to find the page with the listed procedure. She asks Jeanine again, who now knows and responds that they are on step three. Donna continues:

Chapter Eight: Enactment

Before we move onto step three, I want to share with you my, ah, my impressions, as I was going around; I want to share with you some of the things I noticed. I did hear a lot of questions being asked of one another. Sometimes those questions were not specific to what, have someone explain their point of view, but there was a lot of questioning. Not so much why you think so, or how you think that happened, Junior, would you put that down please! Junior, would you put that down please!

SOCIAL SKILL TALLY SHEET

SOCIAL SKILL: Ask questions to help understand one another's point of view.

DIRECTIONS: The COACH will make a tally mark in the space below each time someone in the group uses the social skill.

Tally Area:_____

Total Number of Tallies: _____

Team Members:_____

Communicator
Tracker
Coach

Figure 12. Social skill tally sheet used by the coach

At this time there is a short interruption as she waits for Junior to comply. A student from another class comes to the door to ask a question, but Donna tells her that she can't be disturbed right now. After these two, very brief instances, Donna continues to provide her feedback of what she heard and saw.

> In this group over here I saw that the question asking, it almost got to be a joke. They were asking questions about every picture, and it was kind of funny. And you know what? That's normal when you first start to practice a social skill. Anybody whoever played the piano, the first time you played it felt very, very funny, and gradually with practice it gets to feel easier. And with practice, practicing the social skill, it will feel less false to you, where it won't be such a joke. And its important, at first it is gonna feel kind of funny to ask questions of one another, because you're doing it. And again, I saw that in this group as well.

She then moves over to the "Looks Like/Sounds Like" poster and begins to explain that she saw a number of the "Looks Like" things. For example, she saw Joko standing up and pointing to something on someone else's paper, and she saw Annie doing the same, with numerous examples of pointing and trying to figure things out. She next provides an example of how the social skill was used but explains how the response was not constructive. When one person did ask a clarification question the other person responded "You're wrong. You don't know what you're doing." A number of students gasp when they hear this. Donna responds, "Instead of saying you're wrong, how could you say that in a way that would be softer?" Joko: "You're not right." Donna: "Okay, well 'you're not right' is still pretty strong, its pretty judgmental. What might you say?" In response to another student Donna suggests saying, "How about explaining what you think instead." Shaw recommends: "That's not right, or that's not correct, can I come over and help you?" Donna repeats what he says and adds, as a suggestion, "Well, here's what I think about it. This is what I think might be going on."

Donna then comments on the team skills, suggesting that they need to sit closer so they can use six-inch voices. She also provides her feedback of what they need to work on.

> You need to work on staying together as a team. I saw many examples of students, if they weren't paired at a table with people who weren't considered their friends, in general, they were stretching and making eye contact and talking with people at other groups. That's not staying together as a team. There was also an example, a very flagrant, a very serious example, of a group that was absolutely doing something that had nothing to do with the task, cause they were not doing their jobs. They were reading something else that had nothing to do with the task. Okay? These are things we need to work on.

She finishes her feedback at 2:04 PM, and at this time they are ready to move on to step three and make lists of the interactions in their ecosystem. She explains to them that they may make their lists similar to what they talked about yesterday, but that they need to decide how they will do it so that everyone is involved. To emphasize the point of collective involvement, she explains that she will collect everyone's journal and grade them "on everyone's work, combined." At 2:07 PM, she informs them that they have about five minutes to get started before it is time to go home. However, before they begin, she interjects again, and asks for ideas of how they could divide up the task evenly so that each team member is involved. One suggestion was that each person take responsibility of her system and then combine each of the three systems into one list. After a few more suggestions she lets them get to task, explaining that now they have about four minutes, with six or seven minutes to finish up tomorrow. At 2:15 PM, class ends as the students gather up their belongings to go home.

Lesson #6, Oh Deer, An Introduction and Tuesday, Playing the Game. "Oh Deer" is the second lesson of unit two. *A* "Day in the Life" provided Donna's students the opportunity to integrate the new team skill with science, and this lesson begins to incorporate more directly the science content of the curriculum. The two lessons complement each other and help to form the foundation for understanding how organisms interact with resources that are necessary for survival. "Oh Deer" is an activity that specifically addresses the notion that "the number of living organisms changes as the amount of food, water, shelter, and space changes within an ecosystem" (BSCS, 1992b, p. 99). In role playing, either as deer or a specific resource, the students "will investigate the dynamic balance that exists between the size of a deer herd and the amount of available resources" (p. 99). The description here is how Donna sets the stage for this activity and how she and her students practice the game by modeling the appropriate symbols and procedures.

It's 1:34 PM, and Donna, initiating a conversation with her students, asks, "All right, so ah, you comfortable? You have enough room there? Joe, did you get enough to eat at lunch?" There is a chorus of laughter, and Donna continues by asking Jennet if she had something to drink at lunch. Jennet responds that she did, and Joko pipes in that he had a hamburger. Donna probes further to ask if they are comfortable, are they too cold or too warm? Finally, in explaining her purpose, she states, "Okay, the questions I just asked you, related to resources that all of us might need in order to survive." She writes on the board the following words: Space, Food, Water, and Shelter. Continuing with the introduction, she mentions that each type of question she asked was directly related to the specific resources listed on the board. Donna:

In the same way, animals, just like human beings, animals who live in the wild have the same requirements of resources so that they can survive in their ecosystem. Today and tomorrow we're going to be seeing how animals interact within their ecosystem to compete for these resources. We're going to be playing a game. We're going to be studying about deer.

Donna then takes a few minutes to allow students to tell stories about deer they've seen. She asks probing questions to get more information about their stories, and this session helps the students to build a link between what they are about to do with past memories and experiences. There are so many stories to tell that Donna, to the disappointment of her students, has to cut the conversation short.

Their next task is for the students to follow along in the book as Donna reads the introduction to the game "Oh Deer." When finished with the introduction, she asks them, while still in their home teams of four, to take turns reading and discussing the rules of the game. During this small group session, Donna, with clipboard in hand, circulates throughout the room listening and writing notes about what is happening in the groups. She specifically looks for evidence of the new social skill, and when they are done reviewing the rules, she explains the task of the day.

Just like with most games, it's really hard to understand what to do if you just read through the instructions. Sometimes, in order to learn how to play a game, you have to play it a little bit first, before you find out, before you can figure it out. So what we want to do today is we're going to do a little teeny, tiny game here, in the fishbowl, before we take it outside tomorrow and play the game.

Donna assigns a few students to be the deer and a few more to be the resources. The students at their desk have the task of watching the game to make sure the proper procedures and rules are followed. Before they begin, however, she asks Joe to demonstrate the resource signs to the class. So that all in the class may see him, he slowly spins around as he gives each sign. (Hands over the mouth means water, hands on the belly means food, an "A" frame with arms extended over the head means shelter, and arms outstretched to either side means space.) There is loud laughter after each demonstration, and once he has finished, they play a quick "Simon Says" type game so the rest of the class may practice. Donna shouts out the resource and each student is to give the appropriate sign. Almost like a contest, students laugh as they try to get each sign correct. Donna then begins to pick students to do the "dry run" and to model the game. There are numerous "Ooos" and "Ahhhs" as students raise their

■ Chapter Eight: Enactment   129 ●

hands and volunteer to play. Once roles are assigned Donna asks Jeanine to explain to the class what they (the deer) need to do first. When Jeanine finishes, Donna emphasizes the rule that once a deer decides on a resource it cannot change its mind based on what resources are available. She explains:

> This is science and in science there is no cheating! [You] cannot cheat in science because what you're looking for is what happens. You're looking to observe what happens, but if you change it, or you influence the results of your experiment in some way, then you're not really being honest and you're not really learning from that experience. So there is NO cheating, and if I catch you cheating today or tomorrow, you will sit out. It will also, [it] will be obvious in our results [if you cheat]. You will wonder how I can tell if you were cheating, but I can tell. [She tells me that she does expect them to get caught up in the game and either forget, or not bother to follow the rules exactly, but she makes the emphasis to reduce this.]

Its 2:02 PM, and they spend the next five minutes "walking" through the activity, reviewing each step before they actually play. After each trial, Donna stops the class and clarifies confusions and makes sure they know what to do later. For example, those deer who get a resource take the resource back with them to the other end of the room. That resource (player) now becomes a deer. Those deer who did not get a resource die, and become a resource themselves. Students are laughing and a few comment to their friends, "He's dead" because he didn't get a resource. Donna also clarifies that when they (the deer) run to get their resource they need to keep showing their resource sign. After three trials Donna asks if they now know how to play the game, and the students respond with a chorus of "Yeah!" It's time to go home and as they get ready, Donna, with a slight laugh, again comments, over the increasing volume of chatter, "There is no cheating in science!" On the way out to the school buses, Oliver asks if they can't cheat in science does that mean they can cheat in everything else? An interesting thought to ponder for the next day.

The following day was spent outside playing the game "Oh Deer." Before they went out Donna reviewed the procedures and rules for the game. The game was played on the asphalt courtyard. They did ten rounds with Donna organizing the groups and collecting the data. Each group (deer and resources) was separated by about thirty yards. In the beginning there were fifteen resources and fifteen deer. Groups were formed by counting off by fours. The ones and twos would form one group and the threes and fours would form the other. Initially Donna was yelling and running around to make sure everyone got the directions right. She would run over to the resources and indicate who was a predator and who was a disease, and then run back to the center of the courtyard so everyone could hear her. I began to get involved by picking the person to be the disease and predator on the resource side. Anybody who picked one of these would automatically die. There was some confusion, lots of laughing, a little screaming, and plenty of running. The population didn't seem to fluctuate that much, and Donna thought it was due to "cheating," or that the students were not quite sticking to the game rules. This could be true since it seemed that they didn't completely remember the rules. Donna understood that for the first time this would be a problem.

After ten rounds were completed, with the population of deer recorded after each round, it was time to go back inside. The data were used for discussion, to make interpretations about the deer population, and to graph their results. The graphing session took place the following day, and was unique for one particular reason. It was during this graphing session, and a subsequent one on Valentine's Day, that Donna exhibited a style of direct teaching, a more traditional approach than what she normally displays.

Oh Deer, Graphing. During the transition from labs to science, Donna turns on the overhead and places on it the data table from yesterday's game. At 1:35 PM she rings a bell (another quiet signal) and with her book in hand she moves out from behind her work area. She begins by telling the students that they will spend a few minutes talking about the questions from p. 107. These questions are intended to help them link the results from the activity to real-life situations with deer populations and their quest for resources. While the questions are in the book, Donna reads each one aloud to the students, and for each one they take time to generate ideas for response. Donna begins:

> I'd like us to take a look at what happened yesterday, and maybe talk over some of those discussion questions. I'm going to give you, umm, I wanted to give you some time in your groups yesterday to talk them over, but I think we'll just, for the sake of uh, [interruption]. All right, I'd like you to take just a few minutes to talk generally about those questions. So like question number one.

As she reads the question aloud for the class, she points to the first entry in the data table, number thirteen (Table 3), and asks what happens to the herd when there are plenty of resources in the forest (food, water, shelter, and space). Julian responds that the deer herd increases, and

Donna immediately asks him why? Donna paraphrases his response, "Cause they have babies, because they've got enough, you know, they don't all die off in the winter, they've got enough of their resources. Jessica?" Jessica responds, "There's more because, like, if predators, like maybe the, like the predators [maybe eat something else] and they won't eat the deer." The student feedback continues, and one suggests that maybe the deer are strong enough to fight off the predators or run away.

Donna moves on to question two and asks what happens to the deer herd when there is a *limited* supply of resources? Providing an example from the data, Donna asks, "Like when we had twenty deer, we only had maybe ten resources left. What happened to the deer then? Joe?" Joe explains, and Donna paraphrases, sliding in terminology, "They're in competition for it so some of them would die off." Jennet suggests that they might get sick and die. Maggie comments that without a lot of resources they may not be strong enough to fight off predators.

Moving on to question three, Donna asks what real deer use for food, water, and shelter in a real forest? Students suggest ponds or rivers for water and grass and berries for food. One student mentions that maybe they can eat meat, and Donna comments, "Okay, so if deer are carnivores, that is if they're omnivores they eat both grass and meat, they just feed off something else. Unfortunately, deer are only herbivores, and they only eat plants." One student comments that there will be a rise and fall in the population, with more deer dying in the winter than in summer. Jennet then mentions that they can eat the bark off of trees. Donna asks where she saw that and Jennet states that she saw it on *Bambi*. Many students provide their ideas, for example Rhoda suggests that they move around a lot "because in one of these places it might be winter and there's no, all of the grasses and plants died so they'd move where there's sun and where there's grass and stuff to eat." Donna immediately paraphrases, again infusing terminology. "Okay, so they might migrate either to a warmer place where there's . . . [inaudible] . . . resources, or they might, or they might migrate to, ah, or they migrate to someplace where there's still bark left on the trees."

Year	Number of Deer
1	13
2	18
3	14
4	12
5	18
6	18
7	16
8	10
9	20
10	12

Table 3: Class Data Collected During "Oh Deer"

Asking question four, Donna wants to know in a real forest which deer might get killed first by predators or disease? Maggie suggests the weak ones, and one student responds, to a chorus of laughter, the "juicy ones." Finally, asking question five, Donna wants to know if the size of the deer herd is going to change? There is a choral response "Yes." Donna asks, "Why?" (Joi responds, but I am unable to pick it up.) Donna then wants to know if, other than birth or death, there are ways in which a deer herd might change in size? There is no response, so Donna rephrases her question and asks again for ideas on how the population may get bigger or smaller. After a short pause the students begin to offer suggestions (which neither the tape, nor I, could pick up). Joko responds by saying that some deer may move from one place or another because of living conditions, while others may not move, but "if more and more come there, there'll be a lot." Donna then asks the class what they think about Joko's response, and one student indicated that she was going to say that. Donna comments, "Maybe migration might be something, deer might move in from another area."

It's 1:46 PM and this brief discussion session lasted about ten minutes. While it seemed that the students were beginning to make connections, Donna indicated to me later that it may have lasted too long. With a very short pause Donna immediately moves to the next phase of this session, an introduction to graphing. She asks for them to open their books, then pauses, and turns on the overhead. She then begins to talk about their data from yesterday.

Chapter Eight: Enactment 133

> We took our data and we organized it in this chart here. We have a pretty clear idea of what happened in each round, and today we're going to be using a different way of picturing our data so that we can see more clearly how the population rises and falls. Today we're going to be learning about graphing.

She then asks them to turn to p. 111 in their book, the section titled "How to Graph." She wants them to read the section (pp. 111-116) and to "be prepared at the end of the time to teach me how to graph." While reading, Donna writes a new data table on the chalkboard (Table 4). They will use these data to practice graphing. When they finish reading Donna calls their attention to these data and she begins to "walk" them through the process of graphing. It is 1:52 PM. Donna begins.

> On the board, I've displayed some data from a class, this isn't your class. This is a class that had twenty-eight students when they played the game "Oh Deer." We're going to use this data [interruption], We're going to use this data and graph, ah, we're going to graph this data so that we can get an idea of how to graph, so that then you will then be on your own, graphing the data from our class. Now, what do I need to do first?

Year	Number of Deer
1	14
2	16
3	18
4	20
5	20
6	6
7	12
8	14
9	18
10	10

Table 4: Data Used for Practicing Graphing

This is a more formal whole-group session where the students are to tell Donna the procedures for graphing; the procedures they just read on pages 111-116. After her question, and a seven-second pause, there are no responses. Donna indicates that she'll wait for "more lights to turn on" and waits a total of twenty seconds before she calls on someone. She finally calls Jane to help her out. Jane responds and Donna paraphrases this response.

> They [the textbook] use two words; they said that there was a horizontal and a vertical [she writes both words on the board]. Horizontal and vertical, so I have to draw a horizontal line. Where should I make that? Should I do it right down here at the bottom or should I make it here at the top?

She refers to the grid that is placed on the overhead and gets muffled responses of both top and bottom. She then tells them to do something that the textbook doesn't.

> One of the things that I think that they don't mention, that you really should do, is that you should look how many squares you've got on your graph paper, so that you know how to stretch out the information in your graph.... Do I want to count by fives for each line, or do I count by ones, what do I want to do? [Not waiting for a response] Do I want to count, do I need to count by halves, do I need to count by tens? Let's take a look and see what we've got here.

She continues by asking one-half of the class members to count the number of bottom squares and the other half to count the number of squares on the side. Once they get the numbers (twenty on the side and eighteen along the bottom), Donna then shows them where to draw the horizontal line as well as the vertical line (all being done on the overhead).

With the lines drawn she wants to know what to do next. With no student response she calls on Jamie and waits fourteen seconds with no response. Currently, except for Donna, there is no talking at all in the classroom. When Jamie doesn't respond, she asks another student, who indicates that she has to decide what each side of the graph means.

> What should I put over here [the Y axis]? Okay, the number of deer. And what should go down here [the X axis]? You've got two variables, the year and the number of deer. The number of deer goes along here, what goes down here at the bottom?

One student responds, and Donna indicates "The year can go down there" and writes it on the overhead. Next she asks "Now what? How many deer am I going to need to include on my graph? What's the largest number that I have?" A student indicates that twenty is the largest number, and Donna now wants to know how many spaces they have to work with. "If this starts at zero [pointing to the graph] how many lines do I have here?" They then count, and after a ten-second pause for counting, Donna responds "Okay, so I have about twenty-three different lines. Should I go by ones or should I go by fives, or should I go by what?" Joko

Chapter Eight: Enactment

responds that they should go by ones but doesn't indicate that if they went by fives they would run out of room. Donna responds:

> Or if I went by fives all my data would be like crowded down here in a little tiny corner and my little graph would be like this [indicating real small]. Well what we want to do when we've got this, we want to spread it out over the graph this way so it's real clear what's going on. Okay, so if we went by fives, by the time I got here and at twenty, and this whole top part of the graph, we wouldn't even need it.

They continue and Donna wants to know if they label the line or the space between the line. One student indicates that the line is labeled because they are "making a graph." Donna clarifies by telling them they are making a line graph. They move on to labeling the X axis and Donna asks how many years and how many spaces do they have? Again, she wants to know how to space it out. Should each line be a year, or should each line be two years? All the while she is labeling the graph on the overhead, there are pauses between when she writes and her questions. This is a model for her students. She then wants to know how to plot the data.

> All right, let's plot the first point. In year number one, down here is zero [pointing to the X axis], in year number one we had fourteen deer. How do I plot that? I start, I go over to one, and then what? What would you do, Swan?

Swan explains that she then counts up to where fourteen would be and then makes a line there. Donna: "Okay, then I make just a point [as she does it] right there, so at year one, we have fourteen. How about in year two? What are my numbers?" Students quietly respond "sixteen." Donna continues to show and explain. "So I go over to year two. Start at zero, walk this way first, then up, to sixteen. What about year three?" They continue, as a class, to plot each of the ten points. Donna keeps mentioning that she starts at zero, walks this way first (along the X axis), and then up to the number on the Y axis. Donna shows them, on the overhead, exactly how to do it. Students quietly call out the numbers to plot for each year, and Donna completes the graph on the overhead.

When each data point is plotted, Donna indicates that she'll take the ruler and just connect the dots, and she does so on the overhead. A couple of students are humming aloud as the others wait quietly for her to finish. Thirty seconds later, when the lines are drawn, Donna comments, "We've got a pretty good idea about what was going on there. There's one more thing you need to know. What is this?" The students respond "a graph." Donna: "A graph of what?" In almost choral response "Oh Deer." Donna explains:

> Okay, this is a graph, this could be any [she pauses two seconds for quiet], this could be anything, but we have to have something up here that tells what it is. Any time you make a chart or graph it has to have a title, because I see a bunch of lines going up and down . . . so I could call this class results, for, "Oh Deer."

A student says it could be called "Oh Deer," and Donna labels the graph *Class results for "Oh Deer!"* It is 2:02 PM and Donna begins to analyze the data. She points out the two twenties that are right next to each other and wants the students to tell her how she knows that this class is cheating. There are ooo's and ahh's as hands are raised. After a thirteen-second pause, she asks them to look at the graphed results and tell her "how do I know that this class was cheating while they were playing?" She waits another fourteen seconds (27 total) before asking Oliver what he thinks. He's not quite sure so Donna continues, "This is suspicious to me, they've got twenty deer here, and they've got twenty deer here. How many kids are in this class?" There is a number of responses, one "twenty-two" and the rest "twenty-eight," and Donna confirms that twenty-eight is the correct number. "So if there are twenty-eight deer [she means twenty] how many resources are roaming around at the other end?" The class responds "eight" (they knew what she meant). Donna: "How can twenty deer match up with eight resources and get twenty again?" There is a lot of student talking as they think aloud for the reason. Joko attempts to explain, but Donna has to quiet the class down before he talks. He explains, "there's only twenty-eight, and they've got, so there's only eight resources left. . . . ." Donna interrupts, "So only eight people could match up." She then calls on another student but has to remind the class that only one person speaks at a time. Jennet tries to explain, starting and stopping, and Donna eventually helps her out by saying that maybe the kids in this class changed signs or maybe hooked up with their friends and went back to the deer side (thus, unintentional "cheating," but nonetheless, messed-up results.)

It's 2:06 PM and Donna asks them to copy yesterday's data into their science notebook. They are to use the data to do a graph tonight for homework. A few students respond with mild hoops and hollers. They are getting antsy. As they copy the data there is about a two-minute, semi-quiet seatwork session. When they finish Donna indicates that she will provide the graph paper for them. Students begin to comment that by looking at the data they can tell that someone "cheated." Donna: "Yeah what happens there in [years] five and six? How do you know [that someone cheated]?" Donna then suggests that maybe they got confused about the rules (Table 3).

At 2:09 PM she raises her hand. When they don't respond Donna commands, "Listen up! I'm giving you instructions. Listen up!" She waits another ten seconds before it gets quiet and then tells them that the homework for the night is to use the real class data to do a graph just like they did in class. She also wants them to take the book home and read the section called "Graphing the Data," on pages 108-110. She just wants them to read, not to do the "Wrap Up" on page 110. Finally, at 2:11 PM she uses another quiet technique, the three-clap method, to get their attention, and she has Julian get up in front of the class and explain the homework to them. At this time they collect their belongings and get ready to go home.

This classroom session, from 1:35 PM until about 2:10 PM, was entirely teacher centered. Donna led the discussion, getting input and ideas when she could, and during graphing, she was the center of attention as she modeled on the overhead how to graph. Her expectations from this session were limited. All she wanted was the kids to graph the data and to do a good job of it. If they do this, she'd be happy. Upon finishing "Oh Deer" the class continued with the unit, completing Lesson #7, "Tree Ring Storytelling," and Lesson # 8, "Balanced Ecosystems." Donna deleted the final evaluation lesson and added the "Ecosystem Mobiles." A description of the implementation of this activity follows.

Ecosystem Mobiles, Day Two. An Application of Social Skills and Science Knowledge. While the students are in the transition from labs to science, Donna writes on the board: "Science Journal, something to color with, Scissors, Magazine clippings, Creating ecosystem mobile instructions" (which they received yesterday). In addition, Donna has organized the necessary materials for construction of the mobile. In each plastic tray, she has included two wire coat hangers, three glue sticks, yarn, a manila folder, scissors, a hole punch, and two twist ties. Hanging from a light fixture is a sample mobile, one that she made. Its construction consists of two crossed wire hangers with magazine clippings, of plant, animals, or nonliving resources, hanging from yarn tied to the hangers. For sturdiness, the clippings are glued to sections of a manila folder. Yarn connects each interaction (for example yarn connecting a picture of a cat with a mouse) and 3 x 5 cards are attached between these interactions. Written on the cards is a description of the interaction and its importance for survival.

The transition from labs to science takes longer than expected, and Donna, arms crossed, with a stern look, mentions, "Let me know, class, when you're ready." She acknowledges those who are ready and gives them a personal, "Thank you." When all is quiet she tells them that first they need to review the instructions, and to "remember, this is an evaluation." Since it is an evaluation, she will be looking for evidence of all the social skills introduced so far. She also reminds them that it is especially important that the tracker pay special attention during the activity so that the team keeps on task. The three jobs are the communicator, the manager, and the tracker. To avoid any confusion, and even though each student has a copy of the instructions for "Creating an Ecosystem Mobile," Donna reviews the general process for completing the activity. She explains that the interactions should be connected with yarn. A 3 x 5 card should then hang between the interactions, and on this should be a written explanation that not only identifies the connection but also why it is necessary for survival.

Once managers begin to get the materials, the noise level increases. There is much discussion and even intergroup sharing of material. Jane twice comes over to the group of Julian, Jeanine, and Annie and hands them clippings that they may be able to use in their mobile. Swan and Jennet do the same between their groups. Even Donna delivers clippings between groups, and in commenting to me about these student-to-student interactions, she states, "That's nice. A sense of community."

The small group session began at 1:43 PM, and at 2:02 PM, when it is time to clean up, Donna has difficulty getting their attention. Student engagement is high, and finally at 2:04 PM, she gives them the cleanup instructions. By 2:08 PM the students have completed their organization, and it is time to go home.

After class, in a reflective moment on the day's events Donna mentions that before she could do this type of open-ended activity, she first needed to know ". . . what I had." Specifically, she needed to know what she had in students, their motivation, their organization, their ability to remain focused over an extended period of time, and whether or not they responded positively to her "training" them in the use of the social skills within a group process. Ideally, as the year progresses she would like to do more of this type of activity, ones that are more open-ended, that she sets up, and where she relinquishes some of her "control" so that her students take more responsibility in their learning. In addition, and because

Chapter Eight: Enactment

this is her second year using the curriculum, she can "tinker" with it more so that it meets the needs of her students. For example, Ecosystem Mobiles is not part of the curriculum, but because of the time of year when this unit was done (winter), along with her interest to get the students active, she created this activity as a culmination of unit two. She also sees the initial stages of this activity as an opportunity for her students to practice collecting ideas and sifting through their "data" so that they may make "connections" that are necessary for mobile completion. Thus, when more time is needed so that these connections can be made, Donna has no reservations about allowing the students one extra day.

I must also mention that ecosystem mobile is an activity where I had an influence. In discussing the two end-of-unit activities; the *ELABORATE* lesson, "School-yard Ecosystems," and the *EVALUATE* lesson, "Decisions About Ecosystems and Resources;" Donna kept reiterating that she couldn't do the school yard activity because it was winter, and that the final activity made her uncomfortable because the kids need the preceding lesson as a foundation. Her main reason for discomfort was because the EVALUATE activity required the use of the IDEA method and her students had not had any practice with it during this unit. Considering the issue of practice and evaluation, doing the final activity made her uneasy. She also felt that the final BSCS activity was

> ... kind of contrived. It says "Well we have to practice this decision making model we better do it someplace, let's do it here, and then let's test them on it there." And uh, I think that I could probably do that a little bit better, and in the other units I can work more with that decision making model in the Health unit and the Technology unit later on in the year. (Audiotape of Donna's thoughts about the end-of-unit activities)

She also felt that the final evaluation was too much

> ... pencil and paper and answer the questions, and I have a lot of kids in my class who would probably have a better opportunity to show me what they know about it if they were constructing something. I also have a lot of kids with a lot of talent in the visual arts and design, and they might enjoy it [building the mobile], the creative nature of it. (1/9)

Thus, after my suggestion, she decided to "take your sage advice" and add a lesson that she felt more comfortable in using for an evaluation. This lesson, structured like other BSCS lessons, also reinforces other components of the unit, including the cooperative learning model and allowing them to brainstorm three separate lists of interactions within the ecosystem that they picked. Thus, Donna diverted from the curriculum and created her own end of unit activity.

Ecosystem Mobiles, Day Three. Transition into science begins at 1:18 PM, and similar to yesterday, it takes too long. Donna finally intervenes.

> Class, how do you feel about it taking about five minutes for us to get ready here, don't you think that's rather excessive? I do. I think that's way too long. [ten-second pause] Okay, the goal again, is for you to move quickly and quietly.

She then waits a few more seconds, asking individual students if they are ready to begin. Once it is quiet, Donna explains:

> When you get into your groups today, the task is to complete the construction of the mobile for your ecosystem. All of you have read the instructions, so who can summarize for me what you ecosystem mobile needs to show? [two-second pause] I need to see more lights on.

After Alijendra's response, Donna again paraphrases so that she is clear on what Alijendra said.

> Okay, so using an insect as an example of a living resource, you need to show some kind of connection, to show how it survives in an ecosystem. What are the things that it needs to survive?"

One response, from Alijendra and a few of her classmates, is the necessity of water. Donna acknowledges this and moves on to the next question. "What else are you going to be looking for, what else should it show?" She said it shows connections among living things. Jennet mentions the connection between living and nonliving, and Donna asks a general question, "How might it use those other things as a resource? The living and non-living things interconnected?" Donna continues and asks the class how many examples they need from each of the three categories, how should they be represented once they have the pictures, and what is the procedure for presenting the mobile to the class? Individual students provide the answers, and when it seems that they know what to do, she asks them what her responsibilities are. Betsi responds that she should be looking for "teamwork." Donna comments:

> Okay, I'm looking for things that are evidence of teamwork, and specifically, asking questions and practicing the [social?] skills. I'm also looking for evidence that you guys understand a lot about the ecosystem and the connections, the one that you picked. [a seventeen-second pause to answer a question] Quickly and quietly report to your teams.

■ Chapter Eight: Enactment   141 ●

The transition took thirty-five seconds, after which Donna used another thirty seconds more to make sure they were settled and ready to begin. She then announces that the manager can get up and get the material. It is now 1:29 PM. The class becomes quiet, and Donna moves around, clipboard in hand, responding to student questions. Using the same technique as during labs, when she approaches a group to respond to a question, she randomly asks any group member, "What's the question?" If there is no response, or a delayed response, she moves away. This is her way of keeping all children on task and making sure that they communicate with each other first before they ask her for help.

One group asks if a map and a student studying it would be an interaction necessary for survival. Their justification was that by studying the map the boy would become intelligent, and intelligence is necessary for survival. As long as their argument was sound Donna said it was okay. She is excited by all the discussions going on, and decides to give them one more day to complete the construction. Her excitement is in part due to her students' energetic response to actually making the mobile as well as the dialogue that she hears as her students discuss interactions within their ecosystems. By taking more time, Donna mentions to me that her students will be able to more fully formulate and internalize the connections and understand the importance these connections have to survival.

As it gets closer to cleanup, the pace and excitement level increase. She has yet to tell them that they will have another day, and at 2:04 PM she finally does so. She explains that because they have worked so hard at the task and are still not done, she will give them one more class period to finish up. There are sighs and many sounds of exhaled air as the students are relieved to have another day. Now, more than ever, they seem excited to complete the task, and today, cleanup is a dream. While Donna hands out extra folders to put their clippings in, two boys take turns sweeping the entire floor while others straighten desks and put trays away. Motivation seems high, with the anticipation of completing the task tomorrow. Even Donna is excited, because the group who picked a "Mall" as their ecosystem has finally realized a problem with their pick and has decided to switch to a "Home" environment. Such a change is direct support for her decision to allow more time to complete the task. They need the time to synthesize and internalize their ideas.

During this class period the social interactions were energetic and friendly; Donna smiles a lot, provides gentle physical contact (hand on shoulder, or arm around shoulder) to help support her students. She attempts to build a confidence in them to accept the responsibility of any task that she may ask of them. Within the school year opportunities are many and include responsibility to do the specific class work, positive personal behavior in class as well as respect and decent treatment of others, and doing errands outside of class. All of this is her way of building a community within the class, a community based on self-respect, mutual respect, and a common decency: a universal model of decency and understanding that has, at its foundation, the essential and positive use of social skills.

Math/science Labs:
The Integration of Math and Science
in an Investigative Process

> All right. Any questions? Is there any group that doesn't know who is going to get the lab? Raise your hand if you are going to get the lab. Take a deep breath, be calm, there's a lot to do. *Grab a lab!* (Donna at beginning of a Grab-a-Lab session)

Upon hearing their cue from Donna, the designated student from each group scurries up to the bulletin board to grab a lab folder. Knowing that they got the lab that they wanted, "Fruit Facts," Angie and Jasmin jump up and down, hug each other, and talk excitedly, as if they had just found out that the love of their dreams likes them. Movement is smooth yet rapid as each pair of students finds the appropriate lab station. Initially, the class is quiet as students read the lab instructions. After about two minutes the noise level is on the rise while groups begin to discuss how they will complete the lab. Rulers and pens go to paper as the students diagram in their lab journals the necessary tables for data collection. Labs are completed in pairs, and each lab station has two pairs working simultaneously on the same activity. If questions arise that the students cannot answer, either within their pair, or their group, they "red-cup it," placing a red plastic cup over a green one to indicate to the teacher that they need help. In her movements around the class, the teacher, clipboard in hand, "hovers" over student groups, listens to their conversations, and jots down pertinent information that helps her remember the types of questions (problems and solutions), interactions, and discussions the children have as they complete the lab activity. These notes are used in assessing the students' understanding within the context of the actual activity as well as used to discuss with the class the things that went well during that lab period. Upon completing the lab, the students are expected to write up their results with lab journals due the following day. Motivation is high because those who turn in a completed lab journal on time are given the privilege of choosing first during the next lab session. For these urban sixth grade students, lab day seems to be one of the most anticipated hours of the week.

Lab time is a one-hour, independent investigative process, in which help from the teacher can be minimal. It is a highly structured session with a considerable amount of time spent by Donna in searching for activities, writing them up, collecting (often buying) materials, and organizing the labs in a way that allows the students to work autonomously. During these times the teacher may act as a true "facilitator" or guide to student learning. However, there are numerous instances where direct teaching is warranted, either as a mini-lesson before lab session or in response to a red cup. It is at these moments where a more traditional approach to instruction is seen.

As mentioned at the beginning of this chapter, the written description of labs is different than it is for BSCS. While I observed labs for eleven consecutive sessions, with additional sessions in the second half of the year, the findings do not lend themselves to a chronology of events. However, while lab sessions are intended to provide the students with independent investigations, Donna does get involved, and there are two general patterns of that involvement: Guided learning and Direct teaching. Presented here are descriptions of the lab sessions based on these categories. The first section will show how Donna uses mini-lessons, in a whole group format, to provide guided and direct teaching. The second section will describe instances of individualized and group help as she moves around during labs. As with the mini-lessons there are times of guided learning as well as direct teaching.

Mini-Lessons: A Whole Class Session

Guided Learning. Donna uses her notes and student feedback to assess her students' understanding of labs as well as to decide when she needs to intervene and when to probe the students with guiding questions. When she feels the whole group can benefit from her feedback, she will begin lab session with a five-to-ten minute mini-lesson. Often, these lessons are focused on the organizational process of the labs and not on the answers or solutions themselves.

For example, very early in the lab process Donna's student teacher began the lab period with a mini-lesson on how to write more detailed lab procedures. Lab write-up question number two, "What did I do, step-by-step, to find the answer?", is intended for students to write a complete and clear description of what they did during that lab. However, students seem to have difficulty with this, so the mini-lesson was designed to give them practice.

The mini-lesson began with three written procedures displayed on the overhead. After reading each passage and assessing it on clarity, students assign the description a score of one to five, with five being the most clear and descriptive (see Figure 13).

The students were then asked to evaluate which of the procedures helped them to understand what the person did during the investigation. As a short discussion ensued, the students provided feedback about the clarity of the samples and which one they preferred. They were then asked to pick one of the written procedures, one that they felt needed more clarity, and re-write it in a more detailed manner. Upon completion the students shared how they re-wrote the procedure they chose. After ten minutes of this mini-lesson, the lab session began.

Direct Teaching. Donna uses direct teaching for a mini-lesson to help clarify procedural complications the students may have. For example, in the lab "Diagonally Speaking" the goal of the activity is to draw four polygons: a triangle, a square, a pentagon, and a hexagon. When each shape is drawn, the corners on each polygon are consecutively labeled with letters, beginning with the letter "A." For example, the triangle should have three labeled corners; "A," "B," and "C," and the square will have four. The task is to figure out how many diagonal lines each polygon has, and if there is a relationship between the number of sides and the number of diagonal lines. The students do this by first drawing diagonal lines from each corner of the polygon to all others within that shape, and then tallying up the number of diagonal lines they have.

[2.] First we read the instructions carefully. Then we copied the graphs on the paper. Then we predicted the flavors, colors, how many pieces, calories a piece, grams per piece, and my favorite flavor. Then we wrote down what color went with what flavor and we guessed how many of each color would be. Then we opened the pack and put down how many of each color there really was and we checked how close it really was to our prediction. Then we colored each lifesaver on the paper according to how they came out in order from the first to the one we saw last. Then we checked the most and least frequent and if there was only one. Then we checked off the chart and answered the two questions at the end and we did the Lab Response Questions.

■ Chapter Eight: Enactment   145 ●

2. a. I copied the chart.
    b. I unwrapped the roll.
    c. I filled in the chart.
    d. We ate our "Lifesavers."

2. We wrote the colors in order.

Figure 13. Student samples of a written procedure to lab response question #2

Donna noticed that her students were having difficulty keeping track of each diagonal they drew. The task became even more difficult as the number of sides increased because listing the diagonal lines twice was a common error. Therefore, her mini-lesson modeled an organizational procedure that would help her students to keep track of each diagonal that they drew. Following is a description of this mini-lesson.

Donna begins the lesson by drawing a hexagon on the overhead and labels each corner from A-F. Beginning at "A" she asks her students to draw a diagonal to "C," and she does the same. One student then mentions that they should next go to "B." Donna asks, "Are we finished with "A"?" Realizing they are not, they continue drawing from "A." So as not to repeat a diagonal, Donna diagrams a table on the overhead that will help them to keep track of each diagonal they draw (Table 5). When completed with each polygon, the students then create a table to help summarize their final data. Donna provides a sample table with the number of diagonals filled in for a hexagon (Table 6). When finished with the mini-lesson she asks the students to quickly and quietly report the appropriate lab station.

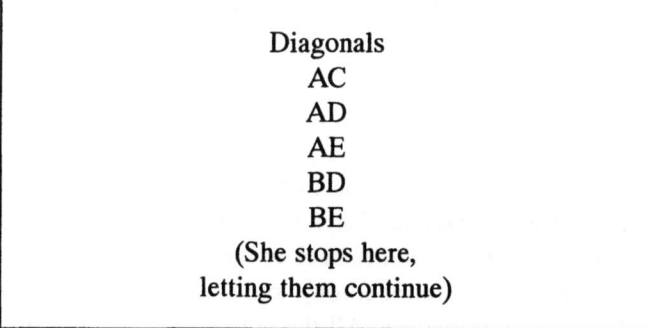

Table 5: Sample Table for Keeping Track of Each Diagonal Drawn

# of sides	# of Diagonals
3	—
4	—
5	—
6	—
9	—

Table 6: Sample Data Table Showing Number of Diagonals per Hexagon

Individual and Group Help

Guided Learning. As she moves around the class during lab session, Donna is often asked questions by her students. Usually, in attempting to get her students to be less dependent on her and to get them to discuss among themselves possible solutions, she will answer a question by posing a question back to the group. For example, during activities that include the use of the pan balance as a tool, where students have Cube-o-Grams (plastic cubes with a mass of 1 gram and a volume of 1 cubic centimeter) and various-sized metal washers as their masses, the students, especially at the beginning of the year, often ask Donna for the mass of the metal washers. Her response, predictable and consistent, is, "How can you find out?" Her intent is to guide the students to learn how they may solve the problem, knowing quite well that all they need to do is mass the washer as they would any other object, by using Cube-o-Grams. This is an example of how she tries to train them in problem solving.

There are also times when Donna uses the final product of the lab as a model for completion of the task. For example, during the building of a tetrahedron out of three triangles made from drinking straws, Donna uses the middle stage, two triangles together, as a model to show how to get from one to two triangles. Holding up the model, Donna asks "What do you need to do to go from one to two?" Jessica figures it out and explains her solution to Donna. When each member in the group has the two straws together, Donna comes back to the group, holding up a model of the finished product (a tetrahedron), and asks how they now get from two triangles to the tetrahedron. All the while, she does not provide the answer but uses a model of each subsequent stage to prod the group along toward solution.

Chapter Eight: Enactment

At other times, to keep a group pressing forward and to ensure accurate results, Donna often models the proper procedure for lab completion. During the activity "Springy Legs," where students compare the ratio of their standing height to their jump height (and telling her students that Michael Jordan jumps 60% of his body height), Donna comes over to one group and shows them how to hold the pen to mark each height on the piece of paper attached to the wall. Holding the pen correctly, to mark both the standing height and jump height, is important for documenting consistent and accurate results.

Direct Teaching. When individuals or groups are stuck with a problem Donna often shows, or even tells, the students what to do so that they may move on to complete the task. The help is not simply probing or guiding the students along but usually is in the form of direct instruction. For example, in September as Donna "hovered" (her term) over one group doing a math lab, she noticed lots of confusion of what was a prime number. After listening to the students' conversation, she finally interjects, "I'm going to come in here because there has been some confusion. What is a prime number?" One student: "I don't know, that's what I asked her" (referring to her partner). With that response, Donna spends the next few minutes telling the group what a prime number is. When finished, and asking for and getting an example of a prime number, she moves on to help another group.

In another math lab, "Orderly Operations" (from Stenmark, Thompson, & Ruth, 1986), the students need to follow the proper sequence, multiply, divide, add, and subtract, in order to complete the problems. Donna shows Omar that if he went straight across the row of numbers, he would get one answer, and if he changed the sequence, following the order of operations, he would get another. She then mentions to him that to remember the proper order, she uses a mnemonic, "My Dear Aunt Sally" (For Multiply, Divide, Add, and Subtract).

In "Going Bananas," students first need to predict whether the banana will sink or float. They then have to estimate the mass of the banana as well as which weighs more, the peel or the pulp. One group's confusion arose with the terms mass and weight. They didn't understand the difference. Donna intervenes and goes on to explain the difference between mass and weight, by telling them that weight relates to how much gravity pulls you down, and mass is more related to size. She continues and tells them that they would have the same mass on the moon, but their weight would be different, because the moon has less gravity.

During the same lab activity, with a different group in a different session, Donna attempts to guide them, but when the group members are still unsure what to do, she steps in and shows them. The problem was that the students did not know the mass of the metal washers. Donna asks, "How could you find out?" She then asks whether they knew the mass of any of the other pieces in front of them. When she receives no response, she tells them that the Cube-o-Grams are one gram apiece. She then shows them how to determine the mass of one washer by putting a washer on one pan of the pan balance and a couple of Cube-o-Grams on the other. She didn't complete the massing for them, but she "hovers" to make sure that when the two pans were even the group knows they are finished. Later, with the same group, Donna gets the same question, but this time they want to know the mass of a small brass washer that Jamie just found in the bag of masses. The problem is that they are in the middle of massing their banana, and without starting over again, they didn't know how to find the mass of this new washer. Seeing that they are close to completing the banana, Donna, not even addressing their question, finishes the massing for them. She adds Cube-o-Grams to the one pan, from 198-211 grams, until the two pans are even. When done, Donna walks away without addressing the question, and Angie and Jamie immediately empty the two pans and mass out the little washer, finding that it is one gram. Finally, Donna shows this group how to write the ratio of the banana to the pulp. First, she asks what ratio is, and Sugie responds that it is the average. With that, Donna shows and tells them how to set up the ratio of pulp to whole banana. She writes 126g:209g on her clipboard and shows the group what she wrote.

Later in the year, during "Fruit Facts," Donna again shows a group how to set up ratios. This time she writes on her clipboard ___:___, and tells the group that this is how ratios are set up, with one line representing the mass of the orange peel and the other the mass of its pulp. After the students fill in the appropriate numbers, she explains to them that ratio is a means of comparing two numbers. In this lab it is a comparison of the mass of the pulp and the peel of the orange.

My final example is in the lab "Patterns and Functions," where Donna offers guided and direct instruction to the same group within the same activity. She guides the group into solving the following pattern: A, B, D, __, __, by directing their attention to the alphabet line located over the back bulletin board. She counts down the line and gives them guidance as to which number comes next. The pattern for this problem was 1+A=B,

■ Chapter Eight: Enactment   149 ●

2+B=D, 3+D=G, and so on. When she walks away the group members continue to use the alphabet line as their guide. While not telling them this pattern, Donna gives them direct guidance on how to solve the problem. However, in another problem she does directly show this group how to figure out the first pattern so they may continue on with the others. The pattern is 4, 12, 8, 12, __, __, __. She tells them, "It's a function," and then indicates that they may either add/subtract, add/multiply, or possibly multiply and then subtract. For this problem she tells them to add first and then subtract, going through the first four known numbers, and two of the unknown numbers with them. By doing so she gets them to start thinking about how to find the patterns.

As indicated, a variety of teaching techniques are used to help the students complete the labs. Donna does not consider that her direct telling and teaching is an imposition. Most often she tells them how they may go about finding a solution, without telling them the exact answer to be found. She mentions that there are times when the teacher's role is to "back off," and there are other times where you need to "stride right in and tell them."

As the year progressed, the number of red cups per lab session were greatly reduced, with only seven being displayed during labs on April 12th, and these were more to show Donna what they had done or to ask for more supplies. Students are also invited to make up labs of their own, especially for science fair. However, science fair, as an activity, while seemingly a combination of BSCS and labs, is actually slightly different than both of these science components. A description of science fair will be given next.

Science Fair: Group Process and Formal Investigation

As mentioned in Chapter four, science fair is where students work in pairs to explore, design, and conduct an experiment of their choice. With one exception, science fair has most of the combined similarities of the BSCS curriculum and the labs. The one component that makes science fair different is the adherence to a more formal structure of scientific investigation, which includes controlling and manipulating variables as well as a more formal format for writing up results: a statement of purpose, a hypothesis, a list of materials, a description of procedure, results, and conclusions. Also, as opposed to BSCS or labs, where there are general and specific topics and goals, science fair is an independent project where students are left to decide the specific direction that they want to take. Throughout the year Donna emphasized that expanding on ideas from labs may be a good way to find a topic worth investigating for science fair. She also stressed, which she clearly outlines in a class handout, "Responsibilities and Requirements for Science Fair" that what the students do for science fair must be an experiment where they are able to collect data and graph and/or tabulate their results.

Science Fair Preparation. Discussion of Variables

It's 9:06 AM and displayed on the overhead are the definitions of two types of variables, independent and dependent, that are associated with scientific investigations (see Figure 14). Donna calls for everyone's attention up front so that they may begin the mini-lesson. "All right, when we talk about the independent variable, [pause] in your experiment that's what you decide to change or manipulate. All right, those are the things that you change. In your experiment, what did you change?" She asks Maggie what she changed in her experiment, but as Maggie talks others are still talking. Donna responds, "I'm sorry, I can't hear Maggie. Please, one person speaks at a time."

Independent Variables	What you changed or manipulated.
Dependent Variables	What the effect was, or what responded to what you changed. What happened as a result?

Figure 14. Overhead list of variable types

She asks another student, "May I have your attention up here?" Finally, after Maggie's description, Donna asks, "So you changed the two different sodas? One of which was Coke. One of which was Seven-Up. All right. What was affected then?" Maggie seems unsure of what she means, so Donna continues to probe, "What was affected? What happened? Were you able to tell the difference, or no?" Maggie says "No." Donna asks, "They weren't able to tell the difference between the two of them?" and continues to probe, asking again if they were able to tell the difference between the two sodas. Finally, Maggie describes what happened, and Donna uses this conversation as an example to explain the meaning of the *independent variable*. Donna explains, "The ability to tell the difference then, was the effect. That's what responded to that change. Right? So your independent variable was?" Donna pauses and waits for Maggie to respond, and when Maggie indicates what the independent and dependent variables are, Donna immediately asks another student to tell what the different variables are for her experiment; "How 'bout in your experiment, Jorge, what were the things that you changed?" After a brief description, Donna paraphrases, "you used different kinds of food products? What was affected then?" As Jorge tries to explain she asks him what they were testing and he indicates "Smell."

Donna: "You also changed something else, you looked at . . . ."
Jorge: "The boys and girls. . . ."
Donna interjects, "Whether boys or girls had that, had a, and that again, that you manipulated."

She moves on and asks Maritza what she changed. Maritza indicates that they changed the shell and the nut. Donna responds, "The shell and the nut, and you tested different varieties of nuts too, right? Okay. And what happened then, what was the effect of it, what were the results?" Maritza explains, and Donna paraphrases to see if she understands. "So the ratio of the mass of the shells of different seeds was affected, or from shells to nuts?" (While this discussion is going on, there is siding by other students. In fact, because of other student talk it is difficult to hear Donna, and particularly difficult to hear student explanations.)

After this dialogue of what variables are in their investigations and before they break off from the whole group discussion, Donna describes the procedure for that period. She tells them that once in their groups, she will come around and check with each team to make sure they are on schedule with conducting the experiment and working through the write-up. Indicating to a few groups that they need to try the experiment again, Donna states, "Some of you need to repeat your experiment. I'd like you to take your experiment out into the hall," where they have some semblance of privacy in which to conduct the experiment. At 9:10 AM, when it's time to get into their groups, Donna states: "Okay, I'm going to be checking for quickly and quietly, we know what that means. I'm going to be timing you." Once said, a few students raise their voices and begin to move. With no comments from Donna, the class as a whole realizes that it is not time to leave, and the students sit back down and get quiet.

Finally, at 9:11 AM, Donna tells them to get into their groups quickly and quietly. The transition is noisy and slow, and Donna stares in disbelief. Once they are in their groups she has to remind them of what quickly and quietly means, and that once they get to their groups they are to sit quietly, look up front, and wait for more instructions. Donna: "Why do I need you to look up to me after you've reported?" Shaw responds that they may need to listen for more instructions. She concurs and tells them that the additional instructions are that she will provide the red and green cups, as in lab, to help organize their question asking. "Do not follow me around the room. I will come to you as you work."

At 9:24 AM, the groups finally begin working. Numerous red cups are displayed, and Donna moves around feverishly trying to attend to all the questions (There are more red cups than usually seen in labs at this time of year). Most questions are about the write-up, and while the experience of labs may pay off, where students attain considerable practice making data tables, a number of the questions deal with data organization and representation. Even with all the experience this year, a couple of groups have to be reminded that it is their responsibility to take charge of this project.

For example, Rhoda, Mariah, and Angie are having difficulty with defining their independent variable. Donna asks if they wrote down the definition from the overhead. Angie indicates that she didn't, and Donna asserts that she will not help them unless they begin to take responsibility for their work. Immediately after this comment though, she begins to probe by asking the group, "What did you change?" For the next three minutes, Donna and the group discuss what they changed, how they did this, and how they may categorize these changes. Donna: "Are you ready to go on from here? Because there are a lot of teams that need help."

For other groups she provides direct guidance on what they may need to do. For example, after looking at Jessica's data, Donna tells her, while pointing to the "Oh Deer" graph on the wall, that she may have to do a bar graph to represent her data. The sample graph is a line graph, but Donna uses it as an example to help Jessica organize her data.

For the next half hour, Donna moves from group to group, provides advice and guidance, quietly directs, and sometimes reprimands those who have ignored their responsibility. She simply tries to maintain the flow of the class and keep them on task to finish the projects. The cups work well as no one follows her around to ask questions. This type of organization of a group process is central to Donna's teaching style and has been emphasized throughout the year. She has trained her students in all components of learning - science, social studies, mathematics, language arts, and especially in math/science labs - to work as a team, to ask questions of teammates, and to use the teacher only for clarification when the team has exhausted all other avenues of problem solving.

However, and possibly because of the independent nature of this work, there are numerous offtask behaviors. Maggie puts the cups over her eyes and pretends they're glasses, while Joe, Julian, and Shaw spend considerable time talking and giggling. To this latter group, Donna comes over and tells them, "I am not going to be responsible" [for completion of the experiment]. She then continues, on her knees, at eye level with the students, to explain that they must be responsible for this. Without stopping to hear their reply, of which there is none, she moves directly into a conversation about their independent variable, what changed, and how they can organize the data. Finally, before she leaves, she explains once more what the term "independent variable" means.

Today's session lasts until 10:00 AM, and her pace is quick and deliberate. Red cups are everywhere as students vie for her time. Much of the confusion is on the notion of variables and how to organize these ideas for graphic representation as well as for a clear and coherent write-up. This process, in this independent and formal manner, is somewhat new for the students, and thus they need considerable teacher guidance, more so than in other activities.

Science Fair. Library Research. Today is library research day. Each group is expected to have at least two references in their science fair literature review. At 9:01 AM, before they leave for the library, Donna announces, "For your science fair, all written material should be done by Friday. Each of you will type a science fair report." She then mentions that for their review she has samples of reports written by former students. Referring to the "Checklist for Science Fair Project Handout" she begins to explain the format of the final written report in (See Figure 15)

As she explains each step, she paraphrases, in more familiar language, what she means for each of the categories. For this purpose, she states, "What you were trying to find out." For the procedure she explains that it "is what you did step-by-step" (this is the same language used in labs). The results are "What you found out," and the conclusions are an explanation of "Why what happened, happened." She emphasizes that they must have examples that support the statements they make, and at the end of the report they must have a bibliography that provides a list of all the references used.

Abstract
Title Page
Table of Contents
Acknowledgments
Purpose (What you were trying to find out)
Review of Literature (Each grp. will review two articles)
List of Materials
Your procedure (What you did step by step)
Results (What you found out)
Conclusions   (Why what happened, happened. Give examples: "Why are boys better predictors than girls?"
Bibliography

Figure 15. Written format for science fair

During this process of presentation and explanation, Donna places a greater emphasis than usual on paying attention. She wants complete silence as she explains what is expected of them. At this point, she turns on the overhead and displays examples of literature reviews written by past students. The samples vary, from two short paragraphs to a complete written page. If they wish to, students can review these for ideas on how to write their report.

Once in the library, students are allowed to work on their own, again, using the cups if they have any questions. A few groups go immediately to the on-line computer system or to the stacks of encyclopedias. However, the same group, Joe, Julian, and Shaw, uses this time to talk, laugh, and giggle. Donna has to intervene again to attempt to get them on task. In addition, Rhoda is told that if she doesn't work she'll have to leave. Donna continues to reiterate that they must take responsibility for this project; she will not do it for them. Thus, while a number of groups are working and making the attempt to find information about their topic, the same students, especially the one group of males, continue to be a challenge.

Science Fair Work. At 9:15 AM, after the weekly spelling test, Donna raises her hand to get the class's attention. When she does, she begins:

> Today your science, your rough draft for your science fair paper is due in your folder. You should have your results completed, and all of the writing would be due in rough draft form. I need to review those folders, all of those materials inside of it, this weekend, to edit it and possibly make some suggestions about ways that you can improve on it, or things that you might want to reconsider, or to change, or just even to comment about what really looks well written. All

Chapter Eight: Enactment 155

right? In your folder today, you should have that outline that I gave you, which includes your purpose, I'm sorry there's another thing that's on here, there's a statement of the variables. Your purpose, procedure, variables. Your results and conclusions will both, your results will show a data table and a graph, but there could also be, your results could also be written up, like we worked with Sugie and Swan to, so they that they stated in sentence form, Rebecca [to get her attention], what their results were. Okay, you could even have that to, so it's the specific results or the what's happened of your experiment. You should tell me in a paragraph what happened. Also, in addition, it's nice to include acknowledgments. What are acknowledgments? When you acknowledge someone what do you do?

One student mentions that an acknowledgment is when you thank someone. Donna expands on that and states:

Okay, so you might thank someone for helping you. Like if Mrs. Levin helped you with something in the library, or somebody gave you an idea for a project, or you might want to thank the people who participated, when you collected the, when you collected data from another class. Thank the teachers who allowed you to use their um, to interrupt their classes and borrow their students to collect data. Also, the review of the literature. That's your summary of the two articles that you found, that helped you to complete the research, or to learn more about what happened, and why it happened, and more of the scientific basis for your experiment. These sections should be in here. All of this stuff, [pointing to the list on the overhead. see Figure 16] is on that outline that I gave you, but these things should be included also. All right? Are there any questions about what needs to be your folder when I collect them at ten o'clock today? [One "No" response and one yawn]. All right, Joko, will you get the lights for me?

As the lights go on talking increases, but Donna cuts them off, indicating that she has more instructions to give. She tells them that they again will use the cups for asking questions. She reiterates that she does need to collect these folders so that she may review the material this weekend, because next week they will be going to the computer lab to type up their final draft. When she is done with these instructions, she asks them to move to their groups quickly and quietly, and she times them. The noise level is high, and eleven-seconds after she tells them to go she yells "Stop! (Except for outside when playing "Oh Deer" or conducting an activity where she must raise her voice, this is the only time all year that I heard Donna yell). Junior, what does quickly and quietly mean?" He mentions they should move without talking. Donna: "Without what? [Junior says "talking"] Right. Quickly and quietly report to your science fair groups." This transition is much quieter and the class structure dissolves into small group work.

During this small group work, Donna filters throughout the class to see who needs help and who has questions. The cup system works well, but there are more questions than in lab, and the time that she spends with each group is considerably longer, a couple of minutes or more. As she moves around, one group begins to tap the cups on the desk. This catches on, and three or four groups copy the first group and do the same. Donna ignores this, and it lasts only about thirty seconds. Junior and Joko drop their cups on the floor, pick them up, and wander around the class, seemingly not knowing what to do.

Variables
Purpose
Results
Conclusions
Data Table
Graph
Acknowledgments
Review of Lit.

Figure 16. List of material to be included in the science fair folder

Donna moves over to Alijendra, Jasmin, Maggie, and Lueere, bends over at the waist, and briefly chats with them about their project. There are at least two other red cups. She now moves over to Joe, Shaw, and Julian, who have the cups separated and spend the time talking and laughing. Donna comes over, grabs the green cup, puts it over the red cup and asks sternly, "Boys, what does this mean? Shaw laughs, but as Donna scolds the group, he turns away from her. This is a familiar pattern with this group, and part of what Donna says is "I don't have time. . . . "

She next moves over to Jamie and Annie and states "I apologize for taking so long. What's the question?" She then gives them graph paper and mentions "You know what? I will come right back to you. Let me scoot around and see what the other questions are." After a few minutes she comes back. "Okay, let's look at your data." Donna then proceeds to help them organize their data by categorizing their variables. They first list the type of beans and then begin to talk about what they measured. Donna lists length, mass, time and talks with them about how to structure their graph. She writes directly on their graph paper and shows them what to put on the two axes. She also shows them how to scale the graph. The class noise level increases, but Donna continues with this group and indicates that a bar graph may be good. "Does that help you? I don't know

Chapter Eight: Enactment

what your data is." Both girls give an affirming nod that it did help, and Donna moves off to help others. The four minutes that she spent with this group is long compared to labs, where one minute may be the maximum.

Donna goes over to Jose and Omar, who have been waiting for a while, and asks what their question is. Jose responds, "We don't know how to do the graph." With a slight chuckle, Donna bows her head and asks Jamie and Annie if they can help them until she comes back. Donna comments, "There's only one of me." About one minute later Donna comes back and begins to look at their data. She suggests that they "take one more step and find the average. How do you find the average? Do you remember?" Jose: "Multiply?" Donna responds, "I think you may be confused. Shall I tell you?" Donna then explains to them what they need to do. She tells them to add up all the boys and all the girls and then divide by that number. At that moment, Oliver walks up to ask a question and Donna states "If you have a question you'll have to put your cup up and wait your turn." Donna turns back to Omar and Jose and labels the axes for them and talks about plotting the data. Next, by asking a series of questions she reviews with them what they have done. "What are you going to do first? How are you going to find the answer? Are you all set? Can I go on to another group?" (Both students are so soft spoken that their responses were inaudible).

She moves back over to Julian, Joe, and Shaw. This group takes a lot of her time for not getting much done. At 9:51 AM she announces that they have three minutes to wrap up. At 9:53 AM, Julian, Joe, and Shaw finally have a question. At 9:55 AM she makes the final round through the class to see how the groups are doing. At 9:58 AM she tells them she will collect their science fair folders and that they need to have out their reading book. At 10:00 AM Donna begins the Status of the Class (a technique where she asks each student what book they are reading), and reading session begins.

Science Fair Day, After the Judging. Donna asked me to be a judge in the science fair. However, so that there was no conflict of interest, I judged the fifth grade students. From other judges I heard positive remarks regarding Donna's students. The judges commented that in discussing their projects most students were reflective and self-critical. Also, when asked simply to explain what they did, most students could easily provide a detailed description of what they did and why they did it.

At 12:38 PM the judging is over and we are all back in Donna's class. The setting is relaxed. Students are quietly talking, moving around, cleaning desks, and asking Donna questions about the judging. Everybody, and especially Donna, seem calmed by the fact that science fair is over. Donna spends a few minutes to review with the class the positive comments she received from the judges. The judges indicated that the students seemed to know what they did, they knew how they did it, and they genuinely liked their projects. Donna also tells them that the judges were "especially impressed with the way you answered the last questions," which asked the students to explain what they might do differently if they could do it again. Donna tells them that "they were impressed with how critical you were of your own work. To me that shows more understanding of what the experiment was, because you need to know the experiment in order to know what to do differently." Jessica and Joi then ask who won. Donna indicates that she only has the raw data and she needs to tally this up. Joko tells her she needs to "cook" the data don't make it raw.

For the next few minutes the students are allowed to relax and clean their desks out. Later they will go next door and watch a video. This provides me an opportunity to converse with a few students. Maggie comes over to her drawer to put away a few books, one of which is the science book (BSCS). Following is the brief conversation that ensued (B = Me and M = Maggie).

B: Do you like your science book?
M: No, (Shaking her head). It's too boring.
B: Why's it boring?
M: Too much reading.
B: Then do you like science?
M: I do now! (Meaning not last year).
B: What do you mean now?
M: Ms. N_____ lets you do things.
B: What about before?
M: We used to have to take the book home, read, and answer lots of questions.
B: Now what do you do?
M: We do activities. It's more fun.

Later in the day, I have another conversation (of course brief like this one) with Julian. He is finishing up his science fair poster board. It is too late for the judges, but next Monday night the parents come in to view them, so he is getting ready for that. (Donna would not allow an incomplete project to be presented at the official fair for judging. Julian, of course, was part of the group that seemed to never be on task. I begin to talk to him about science, using my deep probing first question.)

Chapter One: The Dynamic Balance of Teaching

B: Do you like science?
J: Yeah, I got an "A" in it.

I congratulate him and he tells me why he likes science.

J: Ms. N_____, she makes it fun, she makes it like a game.

He explains to me that last year all they ever did was read the book and answer the questions at the end of each section: the check-up questions. He then explains to me why he likes it here.

J: Here we read, but we do more activities, make things, and she takes us places. (last year they only went on one field trip)

I mention that it seems that social studies is a lot like science, and Julian exclaims, "Yeah, I got an 'A' in social studies too."

CHAPTER NINE
Mediation Between Self and Enactment: The Salient Influences on Curriculum Implementation

> The absurdity of a life that may well end before one understands it does not relieve one of the duty (to that self which is inseparable from others) to live it through as bravely and as generously as possible. (Peter Matthiessen, *The Snow Leopard*, 1978, p. 113)

> But, you know, in society at large, what I feel makes us successful human beings is somebody who's you know, confident in themself but also considers and balances that with you know, what other people are doing and makes decisions that are best for themselves but also considers the feelings of others at that time. (Donna, Interview #6)

Building Community

It seems that what is most important for Donna is the building of a learning community, a safe haven for her students where there are opportunities to build self-respect as well as respect for their fellow classmates. Donna talks about "drying the wood" so that a spark of interest may ignite the fire of lifelong learning. She also refers to the process as "building a corral." As Donna explains:

> Well, it's, it's like making a safe place. The only place that this kind of, that this kind of thing can happen is, is in a place where you feel like, "Yeah, it's okay. I'm gonna come here. I'm gonna have an okay time. People aren't gonna bug me too much. I can be who I am and, I can start out where I am, and learn from there." And, I think that that's kind of the "building the corral." It's a like a, it's a safe place, it's a place where we can all be together and it's where I can also kind of monitor, you know, those who are trying to jump the fence. (Interview #6)

Reminiscent of Donna's ideals are the words of Sizer (1992) as he explains his view of schools. While he specifically refers to high schools, the same can be said for all classrooms in all schools. He explains

> Good schools are thoughtful places. The people in them are known. . . . There are quiet places available as well as places for socializing. No one is ridiculed. No one is the servant of another. The work is shared. The entire place is thoughtful: everything in its routine meets a standard of common sense and civility. At such places do adolescents learn about the thoughtful life (p. 128).

In reference to such an environment I think of M. Scott Peck's (1987) description of community-making. In his view, the goals of a true community are to create a safe place where clear communication leads to ideas of group inclusiveness, appreciation of individual differences, acceptance of humility, and a spirit of noncompetitiveness. All of these features enhance the feeling of community. In the classroom this may mean that the teacher creates situations where students are less dependent (or so it seems) on the teacher, and are more willing to take chances, to struggle and learn with each other, and to accept each other's limitations as well as strengths (see Peck, 1987, p. 71), as each child moves forward toward what Vygotsky (1978) considered potential development.

However, the overall question remains; *What does all of this mean for this teacher, in this classroom, during science instruction?* In the following discussion I will attempt to tease out those influences that are most salient upon Donna as she implements innovative science curricula. Initially, what is clearly evident is that her beliefs do "drive" her practice. If she "feels good" about it, then she is comfortable with what she sets up for her students. In addition, Donna does show evidence of a middle ground of teaching, somewhere between the traditional and progressive frameworks. However, it is not a static position, but a dynamic one that seemingly moves back and forth to best accommodate her students. It is also evident that *how* she moves back and forth in the middle ground, and when she does so is not only dependent upon the moment but also upon her view of her students' needs and her comfort level with the material. From this the following questions may arise (they are not all inclusive).

1. What difference is there between the enactment of science fair and the BSCS curriculum and why is there a difference?
2. Why does it happen that in an independent investigative session such as labs, Donna has numerous direct-teaching interactions with her students?
3. Why is the "Oh Deer" graphing session one of the few times where Donna takes on a mode of direct instruction?
4. What is the difference between direct instruction in a true traditional framework and instruction that is teacher oriented but not traditional?

In order to begin teasing out these ideas, however, it first becomes necessary to return to two of the initial models that shaped the direction of this study: The conceptual model of enactment (Figure 2) and the conception of three frameworks of enactment (Figure 3). Reviewing and reevaluating the models will help to provide a foundation for further discussion and inquiry and will allow us to delve more specifically into the main features of Donna's teaching and how these may relate to the specific components of each model.

Returning to the Source: Reevaluating the Initial Models

Donna shows movement between the traditional and progressive frameworks of teaching science. Her general beliefs regarding student learning indicate a location somewhere right of center, closer to the progressive. However, her feeling of responsibility toward the children and her indication that she is the adult and that she is in charge suggest a position to the left of center, toward the traditional. Thus, by her beliefs and her actions, and depending on the situation at hand, she oscillates back and forth in the middle ground. This is reminiscent of the theme of the BSCS sixth grade curriculum, *Dynamic Balance,* in which Donna is in a continuous motion between these two frameworks of teaching, balancing what she believes is important for her students with what she knows they must learn. The importance here is not to pinpoint an exact position of where she may be in this middle ground but to look at her enactment and how it may be influenced by her general and specific beliefs of teaching as well as her views of the curriculum and the context.

The enactment model presented in Chapter two is generic enough to be used in most educational contexts. Relative to a specific setting, the model can be altered to fit the needs of the study. Upon review of the findings here it seems necessary to modify slightly this model. What Donna does in the classroom is heavily influenced by her beliefs, but how she enacts the curriculum and sets up learning experiences is based on the immediate context of the classroom. For example, the BSCS curriculum is highly structured with group tasks and cooperative learning roles built directly into the curriculum framework. How this is implemented and enacted in the classroom is, however, dependent upon Donna and what she believes is important for her students, and thus, if changes are needed she will make them. The revised model is a representation of how the curriculum and context are actually filtered through her belief system before there are direct influences on enactment. The revised model also more closely depicts what happens in this classroom, and because we now assume a central role in teacher beliefs, it allows us to more easily discuss the relationships between enactment and the curriculum and context (see Figure 17).

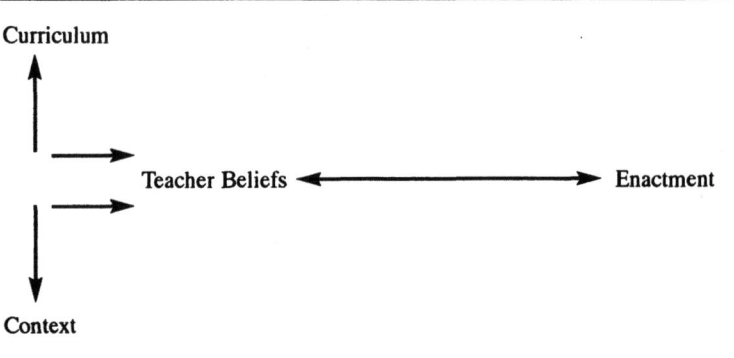

Figure 17. Revised model of enactment

The following discussion will trace backward, from enactment and through teacher beliefs, the impact that the curriculum and the context have on Donna's practice. In addition, this discussion will also bring forth two emerging themes, social skill use and teacher modeling and attempt to place them in the associated component of the model. I have mentioned throughout this study the notion of "messiness," for instance, when trying to separate general beliefs from beliefs in science. The same applies here as I make attempts to categorize neatly what happens in this classroom. I raise the issue again because this classroom is a dynamic environment, where what the teacher does cannot necessarily be "pigeon-holed" into one category and where overlap between components exists.

An Enactment of Science: A Dynamic Balance Among Teacher Beliefs, Curriculum, and Context

Enactment and Its Relationship to Teacher Beliefs

As evident in Chapter seven, Donna believes that what is most important for her students, in terms of science, is learning a process of investigation, and knowing that there is a structure to the discipline. If they can learn how to ask questions, how to find potential solutions to these questions, and how to organize this information to base their decisions on, they are then building a foundation for life-long learning. (Content, she believes, comes along the way in the later grades.) Embedded within this process is also the need to learn how to work with each other, thus, the emphasis on social skills. For Donna it "made sense" to teach the children how to use these skills and to provide them direct feedback as to their use.

Linked to her beliefs of the process of learning, investigating, and problem solving is how Donna goes about providing her students models to work with. Donna constantly models to her students. She provides feedback of "what went well" in activities, and in doing so she models the technique of constructive criticism. In addition, during debriefing sessions she provides her students with examples of appropriate use of a specific skill (usually a social skill). These examples may be real, taken from her observations of student work or made-up to show the students how the use of a particular skill may look like, or sound like. She will also model the delicate task of providing feedback on what the students need to work on and will often suggest ways in which the students can learn to temper their criticism. Thus, she models by example.

Donna will also model her thought processes of how she went about solving a problem or even how she constructed something for the class. At other times she will just show the students a model of a completed project and then ask them to structure it in a similar manner. One example of this latter instance was during labs, when she used each successive stage of building the tetrahedron to prod her students along toward completion. An example of the former was the modeling of her thought processes of how she constructed her "Ecosystem Mobile." Here, Donna took the opportunity to show her students the one that she had made, and walked them through her thinking of why she did what she did. She also modeled to the students how they could make links between this activity and others that came before. Donna explains:

> Today will be the last day for construction for your mobile. A number of the groups, when I was going around yesterday, you had said to me that you, uh, you were having some trouble kind of visualizing how you were going to put it together. So I thought I would give you some suggestions about what I did when I put it together. You know what? I had the same problems that you were having.
>
> The first thing I did when I put it together was I strung the different components, the living and non-living, plants and animals components, all [inaudible], all of these. So I tied a string on there and I put them on, in sort of an up and down sort of way. Then, of course you have outlined what your connections are, that show how the different components interact with one another and guarantee their survival. So then I decided that I would poke other holes and find ways to connect along, sort of like the "Web of Life" [In reference to a previous activity] where we tossed from person to person, we tossed that ball of yarn, you guys remember that activity? Annie, you weren't here. So instead of tossing the ball of yarn from [inaudible] from one thing to another [pause] Omar? [to get his attention] we are going to attach the connections, you might want to either do that between the two, or also just maybe hang something off of there to show,

and say how those connections are made. Does that help you a little more, Maggie, and how you make it? Joe, does that help you a little on how to construct, construct this?

All right, today is the last day, you must be completed, and you need to be prepared to present on Tuesday. Quickly and quietly report to your groups. Get your manager, and get your materials and get started.

On the very next day Donna used the student mobile presentations as an opportunity to model for her students the use of the unit's social skill: "Ask questions to help you understand one another's point of view." For example, an interaction necessary for survival, described by one group, was a little boy and a bookbag. Their reasoning was that the bookbag was needed for learning, which would allow the boy (when he gets older) to get a job, which would bring in money, and allow the boy to buy food and pay for rent (shelter). This group used the need for learning as a necessity to get money, to buy shelter and food, so that the boy could survive. Donna mentioned that these interactions were more indirect, with the notion that learning would eventually lead to getting the resources necessary for survival and explained that this is counter to the typical direct interactions; for example, a cat eating a mouse. Donna continued though and asked for more of an explanation of the role that "intelligence" plays for survival. After Alijendra went through a long explanation, Donna then paraphrased her understanding of the connection, and thus, while not quite modeling question asking, she does explain what she thinks they meant.

So the more removed it becomes from nature, if he were just going to go out and capture an animal and use it for food, it might not be that important [the idea of "intelligence" or formal learning], but in the environment that he is, he lives in, that's where adaptation, the way that he's adjusted to that environment, helps him to survive in that particular [setting].

After Donna's modeling of question asking, student questioning catches on, and during the second group's presentation, students began to ask questions *to help understand another's point of view*. For example, Alijendra asked how the interaction between a boy and a map were necessary for survival. The group members explained that he needed it to find shelter. Julian wanted to know how the Nordic Trak and the person interacted. The justification was that exercise is needed for a healthy body and to prevent disease. Donna probes, "What would you call that? Would you call that an interaction for food, shelter, space, or water, or is that [a] different [thing] that's important for this particular organism?" The response was, if you're not healthy, then you won't survive, but if you are, your chances of survival are better. Donna then asked the class, "Does that mean we might need to modify our definition of the things that we need. Maybe, it appears we also need regular exercise? That's one of the resources we didn't identify."

Therefore, modeling, showing kids how to be successful is important for Donna, whether it is done directly, as in showing them how to construct the mobile, or indirectly, as in her debrief sessions and the question/answer session of mobile presentations. As Donna explains:

> Sometimes it's direct, sometimes it's not direct, sometimes it's just by example, but sometimes it's "well, here is how I would do it.", "here's a suggestion that I would make. Would you like to see what I would do?" . . . I think it's like it, in its simplest form it's an example on which you can base your behaviors. And so it might be a suggestion of one way to structure your work and when, in one case, it might be me thinking out loud as I do something to show kids the kinds of mental processes that I, that I'll go through, or the kinds of decisions that I make. And I think about the uhm, uhm it's been so long since that activity, but the mobiles, when we did that, you know it's like, "here's what I thought about when I put mine [mobile] together." You know, I thought about the place and I thought about you know, what would be there and I thought about how the things would connect and how which things were living organisms and which were non-living, and of the living ones, which were plants and animals. And so, I talked them through that, that's, I slowed it down and instead of just saying "here's what you're gonna do and you have to do this," I kind of talked them through that. That's an example of modeling.

So for Donna, modeling serves a number of purposes. It shows kids how to do something, using the model as a starting point, but most importantly,

> . . . it provides a way for everybody to be successful. But there's, that there's a way that you can be successful. If you don't know what to do so I'm gonna try to help you out in that way. It's also, I think, I think it's also a way, in terms of modeling and using that kind of talking out loud or how I did this, I think that as you become aware of that you, it's, it's like that thinking about what your thinking about or, or becoming aware of what you do. It, it helps you to learn in that way. I think it, it makes teachers of my kids. And if they slow down and think through that process and become aware, then they become aware of where they may make pitfalls and where the pitfalls are and where, where they may make errors or, and become aware of that and conscious. And so I would hope that it [modeling] would also do that, that it would be uh, a way to improve their thinking by slowing it down and making them conscious of it.

For Donna, modeling is used to show her kids one way to go about learning, one process in which they might be able to organize themselves to learn. She asserts, "It's how I teach and I think it's also a good way to learn." Donna's modeling, done almost intuitively, also mirrors the work by Rubin and Norman (1989), who suggest that modeling approaches help to promote process development and/or logical thinking and achievement of students in science. They assert that before students can be expected to perform certain skills, they should first be allowed to practice and have opportunities to develop these skills necessary to complete the task. Modeling provides students opportunities to imitate a successful strategy (as Donna affirms), and practice allows for refinement or a change of that strategy. The goal is, "If teacher modeling can increase student process skill achievement, formal reasoning abilities may also demonstrate a corresponding increase" (Rubin & Norman, 1989, p. 7). Thus, the models that Donna provides are intended to act as a template in which students can build from or in which they can learn from to build their own.

One other concrete example of modeling displayed by Donna in Chapter eight was the session on graphing of "Oh Deer." Donna used that session as a direct model of how to graph. This, more than any other session, was also an example of direct teaching, a traditional approach to teaching science. However, this session was interactive, and students seemed mentally involved, especially as they tried to figure out who had "cheated." Because graphing is not "covered" well in the curriculum, Donna felt it important to "slow the process down" to help students internalize it better. This will be discussed in more detail in the next section.

Enactment and Its Relationship to the Curriculum

The dominant science curricula used in this classroom was BSCS's *Science for Life and Living*, followed by math/science labs and science fair. As Donna has indicated, when she came to Oppenheimer, which was using BSCS, she was not at odds with the structure and intent of the curriculum. In fact, her earlier training in collaborative problem solving had prepared her quite well for teaching in a manner that focused on cooperative learning and the use of social skills. Thus, her beliefs in the teaching and learning of science closely matched the BSCS framework. While her implementation had a high "fidelity" appearance (See Snyder et al., 1992), Donna continuously evaluated her purposes with those needs of the students and made changes accordingly. She indicates:

> My relationship with the curriculum is changing all the time, and I'm, I'm comfortable with it and uh, on some levels and on other levels, I, I will express myself freely and that's just what I do. You know, I kind of use it as uh, as a guideline cause I think the, essentially I like the focus on one theme, the, many of the activities that they do are, are very engaging for kids, uhm, they learn a lot. You know, and I, I think that uhm, some of that is me but some of that you know, quite a bit of that also is the, the structure of that. And there are a lot of people who've had a lot of experience in education who developed it so. You know but again, it's, the choice is gonna be mine. . . . It's gonna be based on what I see and what kids like and what they're interested in, are. You know it's a pretty good bet though that kids are gonna be interested in like, issues that have to deal with the environment and animals and plant life and you know, just the things that are about them. And that's, that's a pretty solid thing to base it on, kids are pretty much into that.

Based on this choice, I observed two notable changes that Donna made to the curriculum. The first one was "Ecosystem Mobiles," and the second was a complete revamping of unit four, which concentrated on problem solving centered on health issues and communication and decision making. Her reasons for adding the mobiles and deleting the final activities in unit two were discussed chapter eight. The changes of unit four were based on her past experience in which she felt that the unit consisted of too much seatwork, with little (or not enough) interactive activities. Such changes are evident of her evolving relationship with the curriculum.

Social Skills

The most important relationships between enactment and curriculum are again filtered through Donna's belief system. The most obvious was her belief that training her students in the use of social skills would benefit life-long learning. It is one feature in which Donna's beliefs and the structure of BSCS were closely parallel, and it was in this aspect of science where the overt teaching of social skills was most often seen. For example, during unit one, most of the debrief sessions focused on providing positive feedback on how the students had worked together (Refer to chapter eight for the details). Training her students in this process helped to build the social foundation for the remainder of the year. However, it didn't end at Unit 1. Each BSCS unit has a particular skill (referred to as "team skills") that the students focused on using. unit two is "Ask questions to help you understand one another's point of view"; unit three is "Discuss many ideas before selecting one"; and unit four is "State someone else's opinion that is different from your own" (BSCS,

1992b, T24). In addition, the use of these skills carried over into labs and science fair, where students spent a majority of the time working with each other.

In labs, by instituting the "red-cup-it" technique, Donna continued to focus on social skills and collaboration. She expected students to ask for each other's opinion and to use each other as the primary resource for problem solving. She also used the cups during science fair, but in this instance it seemed to play more of a direct role in the management of the classroom instead of an emphasis on social skills. Whatever the intent, the use of the cups not only reinforced the importance of social skills, but it also moved some of the responsibility for learning to the students themselves, which resembles a more progressive approach to teaching and learning. In other classroom sessions, for example, in mathematics and social studies, students were expected to continue practicing the social skills, and in the debrief sessions, as with science, Donna would provide the students with positive examples of how they worked together.

Donna uses the social skills for two main reasons. The primary reason is to help teach students to learn how to be less dependent on her, to be more dependent on their fellow classmates, and to be more individually responsible for their own learning. When they need help, Donna helps them, but she is also steadfast, when using the cups, on adherence to the sequence of social communication. The second reason is organizational. By organizing the learning this way, it most often has the appearance of a highly student-centered setting, especially if you walked in during a lab session. However, to get this type of system working requires a considerable amount of teacher time, mainly behind the scenes, as she plans and organizes the set-up. The cup system also organizes the class so that instead of having thirty students asking questions, she would have only eight or nine, based on the number of groups, students asking questions. However, the interplay between student-centered and teacher-centered is complex and subtle. What appears to be student-centered at first has behind it an enormous amount of direct teacher involvement. What seems teacher-centered oftentimes pulls considerably from the students and what they experience. This issue will be brought up further in the discussion of the middle ground.

In addition to the social skills, the BSCS curriculum has a built-in cooperative learning structure. For every activity, the specific group size (almost always three) and the cooperative roles are written into the curriculum. Thus, just as Donna models for her students, the curriculum is set up to act as a model for the teacher. Those who may be less comfortable with cooperative learning have fewer decisions to make regarding the set-up, and can, therefore, concentrate on the procedure and content of the lesson. For Donna, who is quite comfortable with designing a cooperative structure, this may give her more freedom during the lesson, more freedom to get her kids involved with the discussions and to help get their ideas focused on the task at hand. In this sense, the curriculum does influence enactment because it parallels her beliefs. However, when the curriculum is not structured in such a manner, she changes it to resemble more of a BSCS structure. This is exactly what she did during social studies. Thus, the relationship between curriculum and enactment is close when the curriculum conforms to teacher beliefs. When it doesn't, Donna changes the curriculum to conform to her beliefs. Such a process is support for the revision of the model in which teacher beliefs filter all that she does.

The BSCS curriculum strongly supports Donna and what she believes is important for teaching and enacting of curricula. Using the curriculum has most likely helped her evolve this process so that it now is virtually intuitive. Donna once mentioned that her teaching has become so natural it's just "like breathing." However, the hard work continues as she finds new ways to get her kids engaged.

Labs is a creation of Donna's and is thus the purest example of what a curriculum would look like if she designed it around collaborative problem solving. Science fair is "stuck" somewhere in between. It is imposed on her and her students, but given that, she makes every attempt to organize it her way. However, just the very nature of science fair, a true independent investigation, means that Donna has to change her approach to it. The following discussion will look at this movement within the middle ground as Donna adjusts her teaching to fit the needs of her students.

Enactment and the Middle Ground

A lot of what Donna did in science seemed child-centered. There were numerous opportunities for group work and independent investigations. Walking into her classroom and observing during times of these sessions would give the definite impression that the children were "on their own." In addition, there were numerous instances where the classroom structure seemed teacher-centered, especially during the beginning of units and daily lessons. Upon closer examination, however, through numerous hours of observation, it became evident that there are a variety of approaches employed by Donna as she oscillated between a traditional and a progressive framework.

I'll begin this discussion with how Donna describes herself and her role as the teacher in the classroom (also, see Chapter seven for more on this).

> Obviously I'm not totally a constructivist. I believe that kids, uhm, I do believe that kids construct their meaning based on what they know, but I also think that they're guided along and there's a point where you stick your foot in sometimes and you have to say "Hey, you know . . ." There's a point, there's a point at which you can, where it's appropriate to tell. There are points where it's appropriate to say, and you know, and perhaps it can be proceeded of like "would you like me to offer a suggestion of what I would do?", or but sometimes it's just saying, "oh well you know, that's called, this" And, you know, or "in the past that's how the ancient Greeks did this", or you know, suddenly there's a good moment there just to insert those things, and, and then sometimes it's important just to stop and back up and provide a little bit of background knowledge, you know, that's all just based on your judgment of what kids come to [the class with].

From this, it is evident that Donna feels that there are times to stand back and let the kids work, but there are also times where it is her job to step in and provide suggestions or even "stride right in and tell them" direct information so to allow them to progress further on the topic. However, similar to her other conflicts in teaching, Donna also has a conflict with this aspect. She knows that it is important for her students to struggle and to learn via this struggle, but she also knows that when they get to a "wall" and are stuck, she needs to come in.

> Well, you know, we've talked about with hovering and, and when I'm hovering, I'm looking for that. I'm looking for that, for that moment too where I have to [jump in], you know and that's, that's like a finesse thing, Cause when you interrupt, you interrupt somebody's thinking, too. And so maybe they'd get it on their own but sometimes, you know, [knowing] where to provide that help is . . . . That's a lot of my job.

In this discussion about the middle ground I'll begin with the BSCS curriculum and describe the patterns of teaching that occur and provide some insight as to why they occur. In labs, examples of this oscillation are described in detail in Chapter eight, and thus a revisiting of that seems unnecessary. However, because science fair as well as the "Oh Deer" graphing are more teacher-centered, both components are worth a closer look.

BSCS and the Middle Ground. BSCS, unit two, is the best example of Donna's movement within the middle ground of teaching over an extended period of time (labs are the best example within a limited time frame). The general pattern of the unit is as follows. It begins with two teacher-centered sessions, in which much of the students' time is spent at their seats reading and discussing with no movement to science teams. As they move into the unit, the activities become more engaging, and the students become more involved, and thus, the sessions move away from the teacher-centered format (although not entirely). Finally, with "Ecosystem Mobiles" the format moves to almost an entirely student-centered structure, where they are, or can be, working independently of the teacher.

As described in Chapter eight, during the first few sessions of unit two, Donna and the students spent most of the time reviewing what they did from unit one (remember, there was a one-and-a-half month time lag), as well as laying the foundation for use of the new team skill and the new cooperative role, the coach. During the next session they again spent most of the time (thirty-eight minutes) in their seats, in a whole group format, reviewing the concepts of *living* and *nonliving*. They were to actually do lesson five, "A Day in the Life" (An Engage lesson) during this class session but finally moved to their teams at 2:00 PM, with only eight minutes to begin the activity. During these two sessions Donna seemed to be taking great care to build the foundation needed for the remainder of the unit. However, the students didn't seem so understanding, and at the end of session two, Rebecca exclaimed in disgust "We didn't even do science." Donna was also upset by the fact that they didn't get to the activity and asserted to me that they would get more active tomorrow.

These two classroom sessions represent a tension within Donna's philosophy of teaching. She knows that she wants them more active and less dependent on her, but she also believes that she needs to help them build some sort of foundation, corral, or even the "sticks" or supports by which they can then build their own bridge to aid in the transfer of understanding between their previous experiences and new ones. Ultimately, her goal is to have the class become less teacher-centered as the year moves on. However, in reality, what happens is that the units generally begin with more of an emphasis on traditional practices, reviewing previous experiences, and practicing the skills to be used in the new ones. Toward the end of the unit, especially in the evaluative stage, the class structure moves toward a progressive approach where Donna sets up the framework for learning and then allows the students the freedom to experiment and take chances within that framework.

In between the beginning and the end of the unit, the class structure oscillates between whole group, more teacher-centered learning, and small group student-centered structures. In general, and even including (though not always) the initial foundation and review session completed at the beginning of the unit, the whole group structure is dynamic and actively involves the students as Donna uses questioning to probe for their understanding of a topic. In fact, session one of unit two was dynamic with lots of involvement. The problem occurred when they spent two days in a row in their seats in predominantly whole group structure.

A typical science session would begin with a discussion in which Donna would ask the class what they did the session before. She uses these discussions to check for learning problems, to check understanding, and to use as a springboard into other topics. She uses paraphrasing as a mechanism to get at student understanding, and hopes that her kids will tell her when what she says is not what they meant. She also uses paraphrasing as a means to acknowledge the responses of every student, acknowledgments that are beyond nods of the head or simple neutral statements such as "Okay," or "Good." In addition, her paraphrasing allows her to have conversations with her students and to exhibit to them that she takes seriously what they have to say and share. The sharing time allows students the opportunities to talk of their personal experiences and to find some relationship between what they know and have done with what they are doing now. During these types of sessions, the students are active within the discussion process as they are allowed to speak about ideas and experiences that have meaning for them as opposed to more formal recitations in which students respond to questions with seemingly "correct" answers. For Donna, this is another way to "build a positive classroom culture." Finally, Donna will also use paraphrasing to "sneak" in scientific terminology. She will use the terms in a specific and appropriate context. Ideally, and eventually, the students will also use the terms as they discuss the topics. This occurred, for example, at the end of "Ecosystem Mobiles" when students freely used terms like *producer, herbivore*, and *carnivore* to explain their project.

While these discussions are representative of a more traditional teacher-centered format, on closer examination they fall more in line with a progressive and constructivist framework, where students are allowed to share past experiences and current understanding as Donna attempts to help them make "connections" between now and then. This is also very much in line with Vygotsky's (1978) perspective by which

> The child's intellectual growth is contingent on his mastering of the social means of thought, that is, language (p. 94), [and where] Human learning presupposes a specific social nature and process by which children grow into the intellectual life of those around them (Vygotsky, 1978, p. 88).

Expanding on this, and supporting the importance of a middle ground, which has a specific teacher role in learning, is Rogoff (1984), who explains that

> Novices learn under the guidance of others who support their progress through adjustment of task difficulty and who provide experience in the joint solution of the problem (p. 7).

Thus, not only does Donna provide support and guidance during these discussions as she probes for understanding, but she also asks questions that may extend the students to the limits of understanding. In doing so she adjusts the task difficulty to help her students move toward potential development. Therefore, instead of Donna dominating the organization of the classroom social structure and asking questions with "right" answers, she builds an environment where students become part of the social culture of the classroom, in which they feel that what they have to say will be respected and listened to.

Both the BSCS curriculum and labs seem to best foster the building of a positive social climate in the classroom. The BSCS curriculum brings forth specific social (team) skills that the students practice using, and labs promote a social independence, as well as interdependence, where if taken to full advantage, allows the students to solve problems entirely on their own. It seems that in order to promote this type of learning, where social skills become part of the content, Donna has to take on many roles as the teacher. There are times, especially at the beginning of lessons or even the unit, where she acts as the master, the expert in the field, or as she says, "the adult," and takes responsibility for *telling* her students what to do. She will then often *model* for them what she expects, or they will discuss the expectations for a specific skill or practice to be used. Thus, she is *showing* her students what to do, usually in a way that will allow for "success." She next provides *practice*, and walks around and assesses the students' skill level. She *intervenes*, either directly or by guiding, when she feels it is appropriate, or when she is asked. She *debriefs* with her students to again provide models of "what went well" and to provide guidance on what they need to work on. As they move through the unit, discussions become dynamic, and students tend to spend more time at the activities, which culminate at the end, with an assessment project. Here the students practice all they have learned and apply their skills and knowledge to what Sizer (1985) calls an "exhibition" (p. 226), a demonstration of mastery.

Such a process is similar to what Roth (1991, 1993) refers to as "cognitive apprenticeship," and what Rogoff (1984) describes as the role of the adult or peer, who is an "expert in the activity" (p. 5), in the teaching and learning of children. In following the apprentice line, Donna is the master of what to do, and the students follow her guidance. For example, the master carpenter doesn't let the apprentice first build a house. The master first *instructs, tells,* and *shows* the apprentice what needs to be done. The apprentice, under the watchful eye of the master, may attempt simple tasks that are necessary for the overall framing of a house. Periodically, the master provides feedback of "what went well" and what needs to be worked on, all the while guiding the apprentice toward more challenging and difficult tasks. The ultimate goal for the apprentice is to become as proficient as the master, to work on an equal level, and even to surpass the master.

Thus, Donna wants them to learn science, but she also believes that the process of learning science and the process of doing so in a collaborative nature will best benefit her students for life outside the classroom. The students also like her approach because she not only makes the learning active (and they consider it fun), but she takes time to *explain* to them, to help them understand.

For example, in responding to my questioning if she likes this class, Jessica tells me, "Yeah, it's fun."

Me: What makes it fun?
J: Miss N_____ is a good teacher.
Me: What makes her a good teacher?
J: She explains things to you. Last year, Mr. _____ would just tell you to do page 115 and answer the questions. If you didn't get it, he would say that you need to do it.
Me: Miss N_____ lets you explain to each other.
J: She explains and helps you learn. She makes learning fun.

Jessica's comments are similar to those presented in Chapter eight at the end of science fair where two students explained that they liked the active aspect of the class. I even spoke to Lueere the following year, and as she reflected back to Donna's class, she said that she liked it because Donna helped her learn she would explain things to her to help her understand, and that this year, in seventh grade, she missed that.

Struggles in a Middle Ground. Science fair was an instructional and organizational challenge for Donna and represented a struggle for both her and her students. Science fair was imposed on her, and this imposition made Donna feel that science fair was somewhat contrived as it was not part of the ongoing structure of the classroom. It also took an incredible amount of time and energy, during which other classroom components had to be "shut down." They did not begin early in the year but approached this only a few months away from the deadline. Thus, being uncomfortable with its formal structure, and being constrained by time, Donna's role became more like a traditional teacher, where she had to step in and spend more time telling her students what to do. In addition, because science fair is supposed to be composed of experiments, students had to understand the concept of variables. The BSCS curriculum or labs do not cover this at all, and thus Donna had to teach the notion of variables and do it in a short period of time.

Science fair also proved challenging because unlike BSCS and labs, the experiments were to be independent in nature. Students were to come up with ideas, and with Donna's guidance, they were to conduct the experiment as well as collect, organize, and analyze their data. However, this independent process was also new for the class. As mentioned in Chapter four, most of the science activities in BSCS and in labs most always had a defined end point. While the strength of these activities many times was in the process, as students would share their ideas, all students would come to a similar, if not the same, end. Science fair was different, and it was because of this difference that Donna had to spend more time in direct teaching. She would spend time helping to categorize variables she would actively set-up data tables and actively construct and label graphs for individual groups. The intent was to get all students to complete the project. Because of this seeming lack of teacher-directed activities and experiments, a number of the groups (see Chapter eight) were at a loss of where to begin, where to go, and where to end. Donna had to forcefully prod these groups along toward completion. In spite of her urging, a couple of the individuals did not complete the project on time, and thus they were not entered into the actual fair.

One other event, previously described in Chapter eight, in which Donna spends the full period in a direct teaching mode, is the "Oh Deer" graphing session. Here, Donna spends the entire time in front of the class, using the overhead as a tool and "walking" the class through a procedure on graphing. This period is also a more formal recitation session because the kids read from the book the instructions on how to graph and then were asked to repeat these instructions to Donna. During this type of question/answer session, student involvement is often lower. The reason for this could be because they are more concerned with stating the "correct" answer instead of providing ideas and input that are more personal and more related to their experiences. Still, Donna uses this session as another opportunity to model for her students. They did get involved with the discussion of how to graph, and, therefore, even though they were on one task and in their seats the entire time, they seemed engaged as they tried to figure out who "cheated" and how they could tell.

Donna's direct teaching mode most likely arose during science fair and graphing because these topics are not addressed at all well in other components of the science curricula. While there is graphing during labs, most of it is set up for the students as part of the lab structure. In" Oh Deer," which is the first graphing session in the BSCS curriculum, Donna takes on a direct teaching role, just as she does at the beginning of the units. She carefully constructs a foundation for future learning, and she provides a model so that her students may have something to use for future reference. While direct teaching is not at her high comfort level, she knows that there are times when it is her job, as the teacher, as the adult, to "stride right in" and tell her students what to do.

Enactment and Its Relationship to Context

In my many discussions with Donna, I have come to believe that how she teaches is so forcefully driven by her own beliefs that to do it differently would be difficult. This doesn't mean that she would be unable to adjust to ever-changing students and the possibility of changing grade levels. What it means is how she structures the learning and what she emphasizes as important will be less likely dependent on the setting and more likely to be dependent on what she believes as it relates to the setting. For example, on one Sunday afternoon in June, Donna and I drove by a small school located in an affluent suburban neighborhood. We talked about the stereotypical students that would go to that school. Most likely they would be white, from families with a high socioeconomic status, and from a community that would invest proportionally more money per student than at her current urban school. I asked her if she were in this setting, would she do anything differently? She indicated "No," only that there would be more people teaching like she did.

Therefore, while the context of where she teaches most likely wouldn't affect her practice, the urban setting does increase her vigor as she goes about teaching her students. She believes that all students, regardless of background, are entitled to the same opportunities to find out what they're good at: the same chances to achieve success. As evident through an examination of her beliefs, Donna holds true that all children have a strength that can be tapped. She also understands thoroughly the unnecessary hardship and difficulty that most of her students, all urban children, face in trying to succeed in life. Their overall education is not of the same quality as their suburban counterparts, and because of potentially low adult expectations of "these" students, they also do not get many of the same experiences. Donna also knows very well that as an urban student if you are tracked down (Donna sees first hand what Oakes [1985, 1990] talks about) your chances of succeeding in school are greatly reduced, and, thus, your chances in life are limited. For Donna the context increases her sense of "social justice," where she has to make a difference because so many other teachers may not. She talks about how she fears that what she does in her classroom may be at conflict with what her children will face when they leave her. She sees her students having success and learning all the time in her class, but she knows that a change in teacher attitude, and a change in expectations, may help to reduce her students' self-esteem, a sense of self-worth that she has tried so hard to build. She can't follow them as they leave her class, so she tries to prepare them to be able to learn in a variety of situations. However, she has little control over where they go.

For example, the science teacher at the very next grade (seventh grade) is at philosophical polar opposites to Donna. She is a traditional teacher, one who believes very strongly in the content of science. She is also a teacher who will separate students on perceived ability and will keep "these lower" kids from the engaging activities. This worries Donna, and, thus, she thinks about such issues.

> I might as well be thinking backwards, Cause I don't think, because I really believe that, that the teacher there doesn't understand, or doesn't, isn't really familiar with the curriculum that the kids have come from, so, uhm, that's an issue in our school.

Constructing Social Norms and Demystifying Science

One of the key components of Donna's classroom is in constructing and maintaining a social climate conducive for learning. While this climate is not created solely in the context of science instruction, it is this specific context in which the overt training of social skills is most often seen. The BSCS curriculum, as well as Donna's belief that learning to work together will benefit scientific investigations, has a lot to do with this. Breaking the stereotype that science is carried out by individuals, often males, all alone in a laboratory, playing around with beakers and chemicals is one goal that Donna tries to achieve. Thus, an important point for Donna (as well as the BSCS curriculum) is to emphasize that science is a social activity that can be done by all people, regardless of gender.

Necessary for the creation of such a classroom climate is the creation of classroom norms that will aid in this process. Referred to as "social norms" by Wood, Cobb, and Yackel (1990), Donna and her students carve out the guidelines and expectations for the existence of a collaborative classroom. Wood et al. (1990) found in their work in one classroom during mathematics instruction that "these norms established in the setting of the whole class discussions were also crucial to the development of the norms for cooperation and collaboration that guided the interaction of the children as they worked in pairs" (p. 16). In Donna's class the opposite seems more appropriate. The norms constructed for the collaborative work in their science teams were also the ones that structured the interactions within the whole group process. With these social norms also came roles and responsibilities for the students and the teacher. Donna would emphasize, and often asked her students, what her role was as she walked around and observed the group process. Accentuating her role and responsibilities may, again, provide a model for the students that they (the teacher and the students) are in this process of learning together. Such an approach follows closely with Donna's beliefs.

Construction of social norms and breaking down the stereotypical barriers of what science is and who can partake in it also help to provide all students equal access to science. Traditional interaction in science has required an adherence to what Lemke (1990) calls "stylistic norms." Lemke suggests that the rules of stylistic norms of talking or writing science

> ... are a recipe for dull, alienating language. They mainly serve to create a strong contrast between the language of human experience and the language of science. . . . It artificially and misleadingly makes students and the public imagine that science stands somehow outside the world of human experience, rather than being a specialized part of it. . . . But the language of science seems to . . . exempt science from social processes and real human activity, to oppose its language to the colloquial language of common sense. From this comes much of the "mystique" of science and the mystification of science. (p. 134)

Lemke (1990) doesn't stop there but continues to criticize the adherence to a way of talking and writing that doesn't allow for public, mainly student, access to real science learning.

> The norms of scientific language veto most of the techniques that all good communicators know are necessary for engaging the interest of an audience, helping them identify with a point of view, and getting a point across to them effectively. (p. 134)

Thus, as Donna structures the learning of science in her classroom, she makes all attempts to "demystify" science and to make it accessible for her students. Evidence of this can be seen in her whole group discussions, the questions that she uses for labs, the way she paraphrases in everyday language, and the way she "sneaks" in but doesn't require the use of the language of the discipline. Establishing social norms and the nonadherence to stylistic norms have made science available to all of her students on an equal level.

The Value of Structure: Final Thoughts

One final thought must be brought up as it relates to how Donna implements science curricula. Donna's classroom is highly structured. The daily agenda is broken down into the different disciplines. The specific disciplines are organized around their central requirements, including roles and responsibilities as well as expectations for both the students and the teacher. Time during specific lessons is acutely accounted for: there is little down time. Activities and assignments have specific purposes: there is no busy work. Students know, or should know, what is expected of them (academically and socially) at almost every single moment of the day. Such structure requires an extraordinary amount of time by the teacher. Is such an environment teacher-centered? If looked at solely by the universal organizational structure, the answer would be, "Yes." If looked at by the day-to-day, hour-to-hour goings-on, the answer would be, "Maybe. It's hard to tell. Sometimes it is, and then sometimes. . ."

I raise this issue of structure because it comes back to one of the central themes of this study and is directly related to one of Donna's central beliefs. As the adult, she has the direct responsibility to make decisions about what the initial structure of the classroom should be. Feedback from students, in the form of how they are doing, is taken into account as she alters this structure. An example would be "Ecosystem

Mobiles." However, she is the adult she is the one responsible, and because she is the adult, and presumably the expert, she should know better how to address certain issues, such as how to teach a specific lesson in science. This mirrors the notion of Donna as the master and the students as the apprentices. As the master she knows what should be done, sets up the learning, and then helps the students, in a variety of ways, to learn to understand themselves. She attempts to guide them toward new potential development, all the while tweaking and adjusting the level of challenge to bring them along to more and more sophisticated academic and social performances.

CHAPTER TEN
An Impact on Teaching, or, an Answer to the Question, "So What?"

The aim of the enterprise [educational criticism], like education itself, is normative, not simply descriptive. Thus the toughest test of educational criticism (and it is the same test I would apply to any form of educational research) is, does it contribute to the improvement of education? (Elliot Eisner, 1991, p. 114)

The preceding quote, from *The Enlightened Eye*, appears in the chapter on "Validity in Educational Criticism" (p. 107). Eisner suggests that valid educational research ". . . should contribute to the enhancement of the educational process and through it to the educational enhancement of students" (p. 14). Any educational research that intends to be useful for policy decisions, teacher practices, or student learning, should, therefore, answer to the question of "So What?" However, one of the difficulties in interpretive research, and especially in a single case study such as this, is to portray the value of the findings in a way that readers may use them by making connections to their own experiences and their own practice of teaching. In my own subjective way, I fully admit that the findings in this study have greatly influenced what I believe to be important for the teaching and learning of science. One of my goals, to describe in detail the role of an exemplary teacher in the implementation of science curricula, has been achieved. For me, the findings have influenced positively my own practice of teaching, and, therefore, they have been valid. However, as Eisner suggests, it is important that others find this study useful as they attempt to implement innovative science curricula. Therefore, in the following discussion I will present the more prominent findings of this study and attempt to describe their importance for the practice of teaching.

Beliefs as a Determinant of Practice

As discussed in Chapter three, and portrayed in Chapters seven and eight, a teacher's belief system will drive her practice and what she does in the classroom. Donna is no exception. Her beliefs that students should be actively involved in their learning as they investigate and problem solve, and that students should learn how to work together, should learn how to ask questions of each other, and how to constructively support each other in learning are all evident in the way she implements the science curriculum. Also evident is how she designs or even changes the curriculum to fit her beliefs of what is important for her students.

Addressing teacher beliefs, as Olson (1981) found out, should be an important consideration for curriculum writers as they attempt to design curricula that meet the needs of the students as well as the matter of the disciplines. If beliefs truly do drive practice, and if teaching is truly a "localized" endeavor (M. Foster, personal communication, February 7, 1994; Hawthorne, 1992), then it would seem imperative to help teachers to not only recognize their beliefs, as Ken Tobin does in his metaphor work, but also to help teachers change beliefs to fit "new" curricula and "new" instructional practices. Donna, through her participation in the critical thinking and problem-solving workshop, does just this. However, not all teachers have this internal drive, or such a high self-efficacy. Thus, it may be futile to ask teachers to implement a curriculum without first asking them to address their beliefs, finding out where they fit with the intended curriculum, and then preparing them to innovate and to use associated instructional techniques.

Proper training of science teachers is exactly what Pratt (1981) suggests in his recommendations for "reducing the barriers to the desired elementary science program" (p. 91). He states: "The data indicate that elementary teachers will increase their use of inquiry/process-oriented science lessons when exposed to the appropriate training" (p. 92). He also suggests that researchers conduct case studies of selected programs in order to reveal what "enablers" are doing. As evident from this study, and in support of this statement, Donna has had frequent training in the techniques of inquiry/process-oriented instruction, and her beliefs strongly support such a learning environment.

Thus, one conclusion and implication for practice taken from this study can be stated as follows. If teacher beliefs drive practice, then as educators, we need to address the belief structures of both inservice and preservice teachers as we attempt to train them in the use of "new" science curricula. By neglecting these beliefs, our efforts in educational reform may be in vain. It seems also important to continue to conduct detailed case studies of what is good in the practice of elementary science and to use these as models for teacher training.

A Balancing Act Within a Middle Ground

The Teacher's Role in Science Teaching

Donna exhibits evidence of a middle ground of teaching science. Such a pedagogical position enables her to achieve a balance between her belief that the students should be active in their learning as they investigate science and the reality of a science curriculum that is specified for a school where the decision of what to teach has already been made. Through her teaching, Donna is also able to address the process of science and, to a lesser extent, the content of the discipline. Other key issues that are linked in the middle ground are the reality of her role as the teacher, the importance of training students in the use of social skills, and the importance of structure within the lessons and activities.

Donna's beliefs as a teacher closely match the previous discussion of cognitive apprenticeship, which stipulates an explicit role for the teacher in the teaching of science. Students cannot efficiently construct all preexisting knowledge in science, and thus the teacher's role must move beyond merely setting up a science learning environment and facilitating learning. She must have a direct role in the process of educating. While both the content of science and a process of inquiry are imperative for a comprehensive science education, what is prominent here is exactly what Yager and Lutz (1994) suggest when they state, *"How we teach* is more vital that *what we teach"* (p. 344, emphasis theirs).

The National Science Education Standards (1996) recommend that teachers of science provide students opportunities to engage actively in activities that develop knowledge and understanding. A further recommendation is that "Teachers select science content and adapt and design curricula to meet the particular interests, knowledge, skills, and experiences of students" (p. II-4). Taken alone, however, this latter suggestion is out of touch with the reality of schooling, where in many instances there is already a required curriculum that the teacher must follow. What is taught cannot come entirely from student interests but must be adapted from the textbooks and teacher guides adopted by the school. This then raises a complex issue. *How do teachers use a specific school-adopted curriculum and still attempt to teach from student interests and knowledge?*

In attempting to resolve this conflict, Donna provides us with a favorable model. She is required to use the BSCS curriculum, and while she cannot entirely do as she pleases, she is able to modify how she organizes the learning within the framework of the curriculum. In doing so, she is able to strike a balance between reality (the school curriculum) and her beliefs of how students should learn science. In this specific case it is particularly interesting to note that the BSCS curriculum closely mirrors what Donna believes is important for learning. Thus, her ability to achieve this balance is most likely easier than if the curriculum did not match her beliefs. Donna, fortunately for her students, has the confidence and a strong sense of self, and thus the ability to make changes that will satisfy both ends: the reality of a required curriculum and her beliefs of what is important for learning in science. To emphasize the point that how we teach can be more important than what we teach, I would like to provide another example. As a high school biology teacher in New York State, I had to teach the content of the Regents Biology curriculum. I could not, for any period of time, venture off into topics of student interest. However, how I taught the curriculum did not matter. Both of these instances lend credence to the assertion by Yager and Lutz (1994), and Donna's teaching acts as an exemplary model for how this can be accomplished in a self-contained classroom.

As the adult teacher, Donna maintains that it is her responsibility to make choices of what to teach within the limitations of the curriculum. Thus, as Dewey (1938) suggested, Donna helps to shape her students learning by "environing" the necessary conditions that will lead to growth. Also, as the adult, she

> ... structures and models the appropriate solution to the problem [and] supports or "scaffolds" the child's extension of current skills and knowledge to a higher level of competence (Rogoff & Gardner, 1984, p. 97).

Donna, therefore, mediates between direct telling and discovery learning by assisting her students and modeling successful strategies that will help them learn and understand science.

The implications here may be most useful in teacher preparation, where novice teachers may have yet to solidify their beliefs of how to teach. A constructivist framework is useful as a referent for teaching science (Tobin & Tippins, 1993), but we cannot be hesitant to communicate to beginning teachers that there are instances when direct teaching and telling students information will be acceptable and useful. While a traditional notion of direct teaching is "impossible and fruitless" (Vygotsky, 1989, p. 150), Dewey suggests that "guidance is not external imposition"

(1902/1990) but is necessary at times to help children make connections between past experiences and current learning.

Returning to the National Science Education Standards, I would like to clarify the point that they are not limited to statements of teaching solely to student interests. The authors maintain that at certain points "teachers need to intervene to focus and challenge the students," and they recognize that teachers do struggle with the tension of allowing students opportunities to explore on their own and to guide them to predetermined goals set forth by the curriculum. Hence, the balance that a teacher strikes between the need for content through direct teaching, along with allowing for student-centered learning, is considered the ". . . . enacted curriculum–the planned curriculum as it is modified and shaped by the interactions of students, teachers, materials, and daily life in the classroom" (p. II-10).

Importance of Social Skills and Structure

Harms (1981) suggests that the "socialization goals" of teaching are detrimental as teachers attempt to concentrate on useful ways to teach science. He asserts that teachers have a narrow view of science teaching, and thus all their energies are expended on getting students to adhere to these goals. He lists the socialization goals of the classroom as

> . . . inculcating students with the work ethic, teaching students to learn from a textbook, paying attention to directions or presentations, carrying out assignments, preparation for tests, preparing for next year, observing the mores of the community, respecting authority, competing, cooperating. (p. 118)

Represented here, what Donna does with social skills, can also be considered a socialization process into the discipline of science. *Socialization goals*, as Tobin and Fraser (1991) suggest for exemplary teaching, could be in the eye of the beholder. In this case, while there is no direct evidence on student achievement, "socialization" into science seems to have a positive impact on the students' interest in learning in this classroom.

The importance of social skills exhibited by Donna and the BSCS curriculum is helpful in the framework of cooperative learning. Unfortunately, "cooperative learning," like "facilitating," can be a buzzword in education, with little or no understanding associated with it. The reform-oriented literature of science education cited earlier does call for more student-to-student collaboration or allowing students to participate in the social nature of meaning making. However, what a lot of this literature does not explain is the *how-to* of cooperative learning through models of real teachers in the classroom. Here, Donna's classroom provides a vivid example of how to structure a cooperative setting and how to train students in the use of social skills. For Donna, it makes sense to teach her students how to use these skills in the context of learning. For her, the assumption that they perform group skills was not enough, and thus, she needed to train them in the process. It can be argued that the social skills, therefore, became part of the content of science itself, and just like learning any discipline, the students needed practice at understanding and becoming competent in their use.

Such an approach, a detailed process of how to implement the use of social skills in the learning of science, and how to continuously provide guidance and feedback to the students is another important finding of this study that has significant implications for practice. Preservice teachers may be told to use cooperative learning, yet with little or no actual training in structuring such learning, implementing such techniques can be futile. Using the enactment process in Donna's classroom as one model, preservice, as well as inservice teachers, will better understand how a cooperative learning environment is actually set up in a real classroom. *"Socialization goals"* in this context may have a positive influence on how teachers implement science curriculum.

Another useful finding for the practice of teaching is the carefully planned structure of Donna's classroom. Time, especially in elementary school science instruction, is extremely valuable, where downtime and transitions can eliminate large sections of irretrievable instructional minutes. For Donna, there are two main components for reducing wasted time. First, in the beginning of the year she establishes a routine for transitions between subjects as well as for those breaks within activities. Second, she manages to get and hold her students' interest by maintaining a strictly adhered to time frame for all transitions as well as for the small group work. Leaving little time for students to "wander" is one way of keeping them on task. In fact, from their studies, Tobin and Fraser (1988) assert that

> ... exemplary teachers used management strategies which facilitated sustained student engagement.... [and] ... A distinctive feature of classes of the exemplary teachers was the high level of *managerial efficiency* (p. 370, emphasis added).

They also explain that exemplary teachers who tend to monitor student behavior at a distance and move around the classroom addressing individual and group questions, may have as part of their belief structure the notion of increasing student autonomy and independence. Thus, as Donna holds her students accountable for adhering to these routines and remaining on task, she is helping her students to become more involved in their own learning. Getting students more actively engaged and challenging them to take more responsibility in the learning process is another goal in much of the literature of science educational reform.

Directly related to the earlier discussion about cognitive apprenticeship and the teacher's role in the student learning is the amount of time the Donna spends behind the scenes structuring her classroom. As the adult, Donna feels that she knows best what to present to her students. Thus, while she does use student feedback to adjust the learning, she is the one who is responsible for the academic subject matter and for the classroom organization. Using BSCS as a guide, which does structure tasks and the cooperative set-up, Donna takes charge in structuring time-on-task and adjusts this as she feels appropriate. I think it is important for teachers, especially novices, to understand how much control they may need to assume as they organize student learning. What is also important here is to recognize the balance that Donna achieves between a strictly traditional framework, where she is in total control, and a more progressive framework, in which a teacher may allow a great deal of student-centered experiences. In Donna's structuring, it becomes clear that she is attempting, at times intuitively, to encourage students to develop their fullest potential in learning "science" and sophisticated social skills. There is "method to her madness," a detailed plan as she keeps her students engaged and holds them accountable for their own learning.

Exemplary Teaching

Throughout this study, I refer to Donna as an exemplary teacher. This conclusion is based on information from Donna's peers, the principal, parents of students, my observations, and student comments. The primary intent of this study, however, was not to make a case for exemplary practice but to provide insight into how a teacher, who is considered exemplary, implements innovative science curricula. A few of the basic findings are that Donna's beliefs do guide her practice; that she balances in a middle ground of instruction, incorporating certain characteristics from both traditional and progressive teaching frameworks; that she emphasizes the use of social skills; and that she continually models how individuals acquire better collaborative learning and problem solving skills. Thus, while my intent was not to make a case for exemplary practice, it seems useful to compare what Donna does in the classroom to criteria set forth by theorists as being ideal in the teaching and learning of science.

For example, in the call for *"science for all students,"* there is an emphasis to promote learning through active engagement in hands-on and minds-on activities, to emphasize the social nature of meaning making, to focus on "less is more," to tie assessment more closely to actual instruction, and to consider that the teacher's role is changing from a dispenser of knowledge to one where she guides or "facilitates" learning. All of these qualities are exemplified in Donna's practice. Tobin and Fraser (1991), however, caution us to "resist the temptation to search for particular strategies that might be adopted by most successful teachers" (p. 232), to resist the search for the one way to teach. Instead they recommend that recognizing the diversity among quality teachers may allow us to "learn a great deal" (p. 232) about exemplary teachers as they teach science (see Tobin & Fraser, 1988; and Fraser & Tobin, 1993, for additional work on exemplary practice). They also suggest, as was done here, that

> A more fruitful direction for research might be to examine a well-chosen case (e.g., an exemplary teacher) to find out what he or she does in the classroom, explore why he or she adopts the practices that are identified, and contemplate the implications for that teacher's changing in a given direction. (p. 232)

While teaching is more complicated than simply asking novices to copy the acts and behaviors of exemplary teachers, looking at these teachers allows us access into what can be successful, or "what works" in the classroom. In teaching we can use these cases as a springboard into how each teacher wishes to construct her/his own paradigm of what it means to teach and learn in science.

As discussed under the topic of cognitive apprenticeship, modeling, or "emulating," successful practices in learning science, *is exactly what is called for in the teaching of science.* Emulating the successful techniques of others is seen in a wide array of human activities as we attempt to model the proficiency of those who are better than we are at certain performances. We model the best tennis player, the best skier, the grand master pianist, and even the master carpenter. Modeling exemplary teachers may thus serve as a useful template for novices, a starting point where they are given an image of what is good in the teaching of science. Just as students of science cannot construct their own understanding of all scientific knowledge, novice teachers need help and guidance as they evolve toward their potential development, the goal of exemplary teaching.

Final Thoughts

As I reflect on my two-year involvement with this study, I continue to think about what is important in the teaching and learning of science. I realize that finding one best method is impossible, because teaching and learning are localized, with the influences of the context, curriculum, and teacher beliefs ever changing from classroom to classroom. This study does, however, provide evidence that based upon certain teacher beliefs there are pertinent "guidelines," or conditions to be met, that help set the stage for exemplary teaching of science.

For example, if a teacher believes that learning is a social process and students construct their understanding of science through social interactions with their peers and the teacher, then a cooperative learning environment is the way to teach. In addition, the findings of this study indicate that corresponding with a collaborative framework for learning is a need for more direct teaching of students in the use of cooperative group skills. Time and again, as I sit and observe students at all grade levels attempt to investigate scientific matters, I realize that many of them have had little training in the effective use of social skills. Assuming that students intuitively know how to participate in cooperative group work is not enough. It seems that being unable to work together is a barrier to learning science. Students need to be taught, and the teacher has the responsibility for doing this. By ensuring that her students learn academic social skills, Donna is helping to provide her students with the skills necessary for inquiry, and as Dewey (1910) asserted when he critiqued the state of science teaching in schools, it is inquiry, science as a method, that is useful for the "making of knowledge."

> I mean that science has been taught too much as an accumulation of ready-made material with which students are to be made familiar, not enough as a method of thinking, an attitude of mind, after the pattern of which mental habits are to be transformed. (p. 122)
>
> Only by taking a hand in the making of knowledge, by transferring guess and opinion into belief authorized by inquiry, does one ever get a knowledge of the method of knowing.... [Thus,] ... in the order both of time and of importance, science as method precedes science as subject-matter. (p. 125)

If a teacher, therefore, believes that "science as method" is important for her students then she must also teach her students in effective group process techniques that are useful in learning. These skills include but are not limited to listening to others speak, encouraging others, not insulting others, and asking questions that help to clarify others' ideas and opinions.

This case study also illustrates the need for more inquiries into the practices of "exemplary" teachers. Are the techniques of such teachers entirely localized, or are there similarities and commonalities of practice across classrooms, across grade levels, and across settings? Tobin and Fraser (1988, 1991; also see Fraser & Tobin, 1993) appear to believe so, and the energy to continue this line of inquiry may not only help our understanding of the teaching and learning of science but our classroom practice as well.

There is, however, a great deal of debate about the best way to teach and for students to learn. For example, an article in the *Chicago Tribune* (Witt, 1995, May 14) for instance, states that the low math and reading scores that children achieved in California are due to "New age, new-wave, touchy-feely ways to educate children" (p. 1, Section 1). Opponents to "new" methods say they have pushed out traditional methods, the tried and true techniques that work in teaching, while advocates of the "new teaching" suggest that poor test results are really a reflection of poorly written standardized tests, and that the new theories were never fully implemented and thus were never given a chance. The actual situation, however, may reflect more the statement of one teacher, who asserted that the problem may be with the futile search for some ". . . magic bullet, when a logical person would recognize that the answer is somewhere in the middle. You need multiple approaches" (p. 10, Section 1).

Random eclecticism in teaching, however, is not the answer. Donna, as represented through her beliefs, illustrates exemplary practice and curricular enactment that balance in a middle ground of teaching and learning. By doing so she addresses the need for the academic content of the disciplines along with a process of inquiry, thus incorporating into her teaching traditional and progressive components of science instruction.

For Donna, there is middle ground in her teaching of science. She not only implements the required curriculum but also structures the classroom environment in a way that actively engages her students in a collaborative process that allows many opportunities for them to become

Chapter Ten: An Impact on Teaching?

more directly responsible for their own learning. For Donna and her students, social skills have become an integral component of the science curriculum. She is a reflective teacher she takes chances as she teaches and tries new "things," but she is informed by theory and practice that these chances are well worth taking. Based on her knowledge of instruction, her understanding of how children learn, and her beliefs of what is important in learning, Donna feels that her approach is imperative for learning in science and other disciplines as well.

In this study I have portrayed a teacher who is strongly committed to her beliefs in the teaching and learning of science and who organizes and teaches her students in a collaborative process that she believes is imperative for lifelong inquiry. This case study represents but one model of how innovative curriculum is enacted and how one exemplary teacher accomplishes this. By adding this study to the growing literature of exemplary practice in science teaching, we attain a clearer understanding of how these teachers enact science curricula and the role that their beliefs have in this process. This not only informs our own practice but also how we prepare future teachers in the teaching and learning of science.

REFERENCES

American Association for the Advancement of Science (1989). *Science for all Americans: A project 2061 report on literacy goals in science, mathematics, and technology*. Washington, D.C.: American Association for the Advancement of Science.

———. (1993). *Benchmarks for scientific literacy*. New York: Oxford University Press.

Atkin, J.M., & Karplus, R. (1962). Discovery or invention? *Science Teacher*, 29(5), 45.

Bandura, A. (1977). Self-efficacy: Toward a unifying theory of behavioral change. *Psychological Review*, 84(2), 191–215.

———. (1986). *Social foundations of thought and action. A social cognitive theory*. Englewood Cliffs, NJ: Prentice-Hall.

Banks, J.A. (1988). Ethnicity, class, cognitive, and motivational styles: Research and teaching implications. *The Journal of Negro Education*, 57(4), 452–466.

Bergman, A. B. (1993). Performance assessment for early childhood. What could be more natural? *Science and Children*, 30, (5), 20–22.

Biological Sciences Curriculum Study (1992a). *Science for life and living. Integrating science, technology, and health: Implementation guide*. Dubuque, IA: Kendall/Hunt.

———. (1992b). *Science for life and living. Integrating science, technology, and health: Teacher's edition. Level 6. Balance and decisions*. Dubuque, IA: Kendall/Hunt.

Blades, D.W. (1997). *Procedures of power and change: Foucault and the quest for possibilities in science education*. New York: Peter Lang.

Bogdan, R. C., & Biklen, S. K. (1992). *Qualitative research for education: An introduction to theory and methods*. Needham Heights, MA: Allyn and Bacon.

Brown Alumni Monthly (1994, February). $50 million of Walter Annenburg's $500-million gift to public education goes to Brown's Annenburg National Institute, headed by reformer Ted Sizer. *Brown Alumni Monthly*, 94(5), 12–15.

Bruner, J. S. (1960). *The Process of Education*. Cambridge, MA: Harvard University Press.

Bussis, A., Chittenden, E., & Amarel, M. (1976). *Beyond surface curriculum: An interview study of teachers' understandings*. Boulder, CO: Westview.

Clark, C.M. (1988). Asking the right questions about teacher preparation: Contributions of research on teaching thinking. *Educational Researcher*, 17(2), 5–12.

Clark, C.M, & Peterson, P.L. (1986). Teachers' thought processes. In M. Wittrock (Ed.), *Handbook of research on teaching* (3rd ed.). New York: Macmillan.

Cobb, P. (1994a). Constructivism in mathematics and science education. *Educational Researcher*, 23(7), 4.

———. (1994b). Where is the mind? Constructivist and sociocultural perspectives on mathematical development. *Educational Researcher*, 23(7), 13–20.

Cobern, W.W. (1993). Contextual constructivism: The impact of culture on the learning and teaching of science. In K. Tobin (Ed.), *The practice of constructivism in science education* (pp. 51–69). Washington: American Association for the Advancement of Science.

Connolly, P., & Vilardi, T. (Eds.). (1989). *Writing to learn mathematics and science*. New York: Teachers College Press.

Cosgrove, M., & Osborne, R. (1985). Lesson frameworks for changing children's ideas. In R. Osborne & P. Freyberg (Eds.), *Learning in science. The Implications of children's science* (pp. 101–111). Auckland, New Zealand: Heinemann.

Dewey, J. (1910). Science as subject-matter and as method. *Science*, 31(787), 121–127.

———. (1938). *Experience and education*. New York: Macmillan.

———. (1990). *The school and society. The Child and curriculum*. Chicago: The University of Chicago Press. (Original work published 1902).

Driver, R., Asoko, H., Leach, J., Mortimer, E., & Scott, P. (1994). Constructing scientific knowledge in the classroom. *Educational Researcher*, 23(7), 5–12.

Duschl, R.A. (1985). Science education and philosophy of science; Twenty-five years of mutually exclusive development. *School Science and Mathematics*, 85(7), 541–555.

Edwards, D., & Mercer, N. (1987). *Common knowledge: The development of understanding in the classroom*. London: Routledge.

Eisner, E.W. (1991). *The enlightened eye: Qualitative inquiry and the enhancement of educational practice*. New York: Macmillan.

Elmore, R.E. (1979-1980). Backward mapping: Implementation research policy decisions. *Political Science Quarterly*, 94(4), 601–616.

Erickson, F. (1985). Qualitative methods in research on teaching. In M.C. Wittrock (Ed.)., (3rd. ed.). *Handbook of research on teaching*., (pp. 119–161). New York: Macmillan.

———. (1984). What makes school ethnography "ethnographic"? *Anthropology and Education Quarterly*, 15, 51–66.

Fawcett, G. (1992). Moving the big desk. *Language Arts*, 69, 183–185.

Fraser, B.J. (1994). Research on classroom and school climate. In D.L. Gabel (Ed.), *Handbook of Research on Science Teaching and Learning* (pp. 493–541). New York: Macmillan.

Fraser, B.J., & Tobin, K. (1993). Exemplary science and mathematics teachers. In B.J. Fraser (Ed.), *Research Implications for Science and Mathematics Teachers: Vol. I*. Perth, Australia: National Key Centre for School Science and Mathematics.

Fullan, M.G. (1991). *The new meaning of educational change* (2nd ed.). New York: Teachers College Press.

Glesne, C., & Peshkin, A. (1992). *Becoming Qualitative Researchers: An Introduction*. White Plains, NY: Longman.

Goodlad, J.I., & Oakes, J. (1988, February). We must offer equal access to knowledge. *Educational Leadership*, pp. 16–22.

Goodman, J. (1988). Constructing a practical philosophy of teaching: A study of preservice teachers' professional perspectives. *Teaching and Teacher Education*, 4, 121–137.

Gross, N., Giacquinta, J., & Bernstein, M. (1971). *Implementing organizational innovations: A sociological analysis of planned educational change*. New York: Basic Books.

Harms, N. (1981). Project synthesis: Summary and implications for teachers. In N.C. Harms and R.E. Yager (Eds.)., *What research says to the science teacher* (Vol. 1, pp. 113–127). Washington: National Science Teachers Association.

Harms, N., & Yager, R.E. (Eds). (1981). *What research says to the science teacher* (Vol. 1). Washington: National Science Teachers Association.

Hart, D. (1994). *Authentic assessment. A handbook for educators*. Reading, MA: Addison-Wesley.

Hawthorne, R.K. (1992). *Curriculum in the making: Teacher choice and classroom experience*. New York: Teachers College Press.

Hilliard, A.G., III (1989, January). Teachers and cultural styles in a pluralistic society. *National Education Association*, pp. 65–69.

———. (1992). Behavioral style, culture, and teaching and learning. *The Journal of Negro Education*, 61 (3), 370–377.

Janesick, V. (1977). An ethnographic study of a teacher's classroom perspective. Unpublished doctoral dissertation, Michigan State University, East Lansing.

Johnson, D.W., Johnson, R.T., & Johnson-Holubec, E. (1986). *Circles of learning: Cooperation in the classroom.* Edina: MN. Interaction Book Company.
Johnson, R.T., Johnson, D.W., Scott, L.W., & Ramolae, B.A. (1985). Effects of single-sex and mixed-sex cooperative interaction on science achievement and attitudes, and cross handicap and cross sex relationships. *Journal of Research in Science Teaching,* 22(3), 207–220.
Kagan, D.M. (1992). Implications of research on teacher belief. *Educational Psychologist,* 27(1), 65–90.
Kamii, K. (1985). *Young children reinvent arithmetic: Implications of Piaget's theory.* New York: Teachers College Press.
Kuhn, T. S. (1970). *The structure of scientific revolutions.* (2nd ed.). Chicago: University of Chicago Press.
Lauber, P. (1987) *Tales Mummies Tell.* New York: HarperCollins.
Lawson, A.E., & Renner, J.W. (1975). Piagetian theory and biology teaching. *American Biology Teacher,* 37(6), 336–343.
Lee, C.D. (1992). Literacy, cultural diversity, and instruction. *Education and Urban Society,* 24(2), 279–291.
Lee, G.C. (Ed.). (1961). Crusade against ignorance. *Thomas Jefferson on Education.* New York: Teachers College Press.
Lemke, J. L. (1990). *Talking science: Language, learning, and values.* Norwood, NJ: Ablex.
Linn, M.C., & Hyde, J.S. (1989). Gender, mathematics, and science. *Educational Researcher,* 18(8), 17–27.
Maclean, N. (1992). *Young men and fire: A true story of the Mann Gulch fire.* Chicago: University of Chicago Press.
Matthiessen, P. (1978). *The snow leopard.* New York: Viking.
Miles, M.B., & Huberman, A. M. (1994). *Qualitative data analysis: An expanded sourcebook.* (2nd ed.). Thousand Oaks, CA: Sage.
Morris, W. (Ed.). (1978). *The American heritage dictionary of the English language: New college edition.* Boston: Houghton Mifflin.
Munby, H. (1982). The place of teachers' beliefs in research on teacher thinking and decision making, and an alternative methodology. *Instructional Science,* 11, 201–225.
———. (1984). A qualitative approach to the study of a teacher's beliefs. *Journal of Research in Science Teaching,* 21(1), 27–38.
National Council of Teachers of Mathematics. (1989). *Curriculum and evaluation standards for school mathematics.* Reston, VA: The Council.

References

———. (1991). *Mathematics assessment: Myths, models, good questions, and practical suggestions*. (J. K. Stenmark, Ed.), Reston, VA: National Council of Teachers of Mathematics.

National Research Council (1990). *Fulfilling the promise: Biology education in the nation's schools*. Washington, DC: National Academy Press.

——— (1993a, February). *National science education standards: An enhanced sampler.* Washington, DC: National Research Council.

———. (1993b, July). *National science education standards: July '93 Progress Report*. Washington, DC: National Research Council.

———. (1994a, May). *National science education standards: Discussion Summary*. Washington, DC: National Research Council.

———.(1994b, November). National science education standards: *Draft for response and comment*. Washington, DC: National Academy Press.

———. (1996). *National science education standards*. Washington, DC: National Academy Press.

National Science Teachers Association (1992). *Scope, sequence, coordination. The content core: A guide for curriculum designers*. Washington: Author.

Nespor, J. (1987). The role of beliefs in the practice of teaching. *Journal of Curriculum Studies*, 19, 317–328.

Novak, J. (1991). Clarify with concepts maps: A tool for students and teachers alike. *The Science Teacher*, 58(7), 45–49.

Oakes, J. (1985). *Keeping track: How schools structure inequality*. New Haven, CT: Yale University Press.

———. (1990). Opportunities, achievement, and choice: Women and minority students in science and mathematics. In C. B. Cazden (Ed.), *Review of Research in Education* (pp. 153–222). Washington, DC: American Educational Research Association.

O'Loughlin, M. (1992). Rethinking science education: Beyond Piagetian constructivism toward a sociocultural model of teaching and learning. *Journal of research in science teaching: Special issue: Science curriculum reform*, 29(8), 791–820.

Olson, J. (1981). Teacher influence in the classroom: A context for understanding curriculum translation. *Instructional Science*, 10, 259–275.

Osborne, R., & Freyberg, P. (1985). Roles of the science teacher. In R. Osborne & P. Freyberg (Eds.), *Learning in science. The implications of children's science* (pp. 91–99). Auckland, New Zealand: Heinemann.

Padilla, M.J. (1991). Science activities, process, skills, and thinking. In S.M. Glynn, R.H. Yeany, and B.K. Britton (Eds.), *The psychology of learning science* (pp. 205–217). Hillsdale, NJ: Lawrence Erlbaum Associates.

Pajares, M.F. (1992). Teachers' beliefs and educational research: Cleaning up a messy construct. *Review of Educational Research*, 62(3), 307–332.

Peck, M. S. (1987). *The different drum: Community-making and peace.* New York: Simon and Schuster.

Peshkin, A. (1993). The goodness of qualitative research. *Educational Researcher*, 22(2), 23–29.

Peterson, L., & Fennema, E. (1985). Effective teaching, student engagement in classroom activities, and sex-related differences in learning mathematics. *American Educational Research Journal*, 22, 309–335.

Porter, A.C., & Freeman, D.J. (1986). Professional orientations: An essential domain for teacher testing. *Journal of Negro Education*, 55, 284–292.

Pratt, H. (1981). Science education in the elementary school. In N.C. Harms and R.E. Yager (Eds.), *What research says to the science teacher* (Vol. 1, pp. 73–93). Washington: National Science Teachers Association.

Raizen, S.A., & Michelson, A.M. (1994). The qualities of an effective teacher of science. In S.A. Raizen & A.M. Michelson (Eds.), *The future of science in elementary schools. Educating prospective teachers* (pp. 30–51). San Francisco: Jossey-Bass.

Riggs, I.M., & Enochs, L.G. (1990). Toward the development of an elementary teacher's science teaching efficacy belief instrument. *Science Education*, 74(6), 625–637.

Rogoff, B. (1984). Introduction: Thinking and learning in social context. In B. Rogoff and J. Lave (Eds.), *Everyday cognition: Its development in social context* (pp. 1–8). Cambridge, MA: Harvard University Press.

Rogoff, B., & Gardner, W. (1984). Adult guidance of cognitive development. In B. Rogoff and J. Lave (Eds.), *Everyday cognition: Its development in social context* (pp. 95–116). Cambridge, MA: Harvard University Press.

Rogoff, B., & Lave, J. (Eds.) (1984). *Everyday cognition: Its development in social cognition.* Cambridge, MA: Harvard University Press.

Roth, W-M. (1991, April). "Aspects of cognitive apprenticeship in science teaching." Paper presented at the annual meeting of the National Association for Research in Science Teaching, Lake Geneva, WI. (ERIC Document Reproduction Service No. ED 337 350).

———. (1993). Construction sites: Science labs and classrooms. In K. Tobin (Ed.), *The practice of constructivism in science education* (pp. 145–170). Washington: American Association for the Advancement of Science Press.

Roth, W-M., & Roychoudhury, A. (1993). The development of science process skills in authentic contexts. *Journal of Research in Science Teaching,* 30(2), 127–152.

Rubin, R.L., & Norman, J.T. (1989, April). "A comparison of the effect of a systematic modeling approach and the learning cycle approach on the achievement of integrated science process skills in urban middle school students." Paper presented at the annual meeting of National Association for Research in Science Teaching. San Francisco, CA. (ERIC Document Reproduction Service No. 305 268).

Schaller, G.B. (1993). *The last panda.* Chicago: University of Chicago Press.

Scharmann, L.C. (1991). Teaching angiosperm reproduction by means of the learning cycle. *School Science and Mathematics,* 91(3), 100–104.

Schubert, W. (1989). On the practical value of practical inquiry for teachers and students. *Journal of Thought,* 24 (1&2), 41–74.

Schwab, J. (1978). The practical: A language for curriculum. In I. Westbury & N.J. Wilkof (Eds.), *Science, curriculum, and liberal education* (pp. 287–321). Chicago: University of Chicago Press. (Original work published 1970).

———. (1978). The practical: Translation into curriculum. In I. Westbury & N.J. Wilkof (Eds.), *Science, curriculum, and liberal education* (pp. 365–383). Chicago: University of Chicago Press. (Original work published 1973).

Science Curriculum Improvement Study (1974). *SCIS teacher's handbook.* Berkeley, CA: Regents of the University of California.

Shade, B.J. (1982). Afro-American cognitive style: A variable in school success? *Review of Educational Research,* 52(2), 219–244.

Shymansky, J.A., & Kyle, W.C, Jr. (1992). Establishing a research agenda: Critical issues of science curriculum reform. *Journal of Research in Science Teaching. Special Issue: Science Curriculum Reform,* 29 (8), 749–778.

Sizer, T. R. (1985). *Horace's compromise: The dilemma of the American high school.* Boston: Houghton Mifflin.

———. (1992). *Horace's school: Redesigning the American high school.* Boston: Houghton Mifflin.

Skolnick, J., Langbort, C., & Day, L. (1982). *How to encourage girls in math and science.* Englewood Cliffs, NJ: Prentice-Hall.

Smylie, M.A. (1990). Teacher efficacy at work. In P. Reyes (Ed.), *Teachers and their workplace: Commitment, performance, and productivity.* Newbury Park, CA: Sage.

———. (1992). Curriculum adaptation. In T. Husen & T.N. Postlethwaite (Eds.), *The international encyclopedia of education: Research and studies, supplementary volume one* (2nd ed.). New York: Pergamon.

Snyder, J., Bolin, F., & Zumwalt, K. (1992). Curriculum implementation. In P.W. Jackson (Ed.), *Handbook of research on curriculum* (pp. 402–435). New York: Macmillan.

Spradley, J.P. (1980). *Participant observation.* Orlando, FL: Harcourt Brace Jovanovich.

Stenmark, J.K., Thompson, V., & Ruth, C. (1986). *Family math.* Berkeley, CA: Lawrence Hall of Science.

Tabachnick, B.R., Popkewitz, T.S., & Zeichner, K.M. (1979). Teacher education and the professional perspectives of student teachers. *Interchange,* 10(4), 12–29.

Tabachnick, B.R., & Zeichner, K.M. (1984, November-December). The impact of the student teaching experience on the development of teacher perspectives. *Journal of Teacher Education.*

Tannen, D. (1991). Teacher's classroom strategies should recognize that men and women use language differently. *The Chronicle of Higher Education,* 37(40).

Tobin, K. (1993). Metaphors and images in teaching. In B. Fraser (Ed.), *Research Implications for Science. Vol. I.* Perth, Australia: National Key Centre for School Science and Mathematics.

Tobin, K., Briscoe, C., & Holman, J.R. (1990). Overcoming constraints to effective elementary science teaching. *Science Education,* 74(4), 409–420.

Tobin, K., & Espinet, M. (1989). Impediments to change: Applications of coaching in high-school science teaching. *Journal of Research in Science Teaching,* 26(2), 105–120.

Tobin, K., & Fraser, B.J. (1988). Investigations of exemplary practice in science and mathematics teaching in Western Australia. *Journal of Curriculum Studies,* 20(4), 369–371.

References

———. (1991). Learning from exemplary teachers. In H.C. Waxman & H.J. Walberg (Eds.), *Effective teaching: Current research*. (217–236) Berkeley, CA: McCutchan.

Tobin, K., & Gallagher, J. (1987). What happens in high school science classrooms? *Journal of Curriculum Studies*, 19, 549–560.

Tobin, K., Kahle, J.B., & Fraser, B.J. (1990). *Windows into science classrooms: Problems associated with higher-level cognitive learning.* London: Falmer.

Tobin, K., & Tippins, D. (1993). Constructivism as a referent for teaching and learning. In K.Tobin (Ed.), *The practice of constructivism in science education* (pp. 3–21). Washington: American Association for the Advancement of Science.

Tobin, K., Tippins, D.J., & Gallard, A.J. (1994). Research on instructional strategies for teaching science. In D.L. Gabel (Ed.), *Handbook of research on science teaching and learning* (pp. 45–93). New York: Macmillan.

Tobin, K., & Ulerick, S. (1989, March) *An interpretation of high school science teaching based on metaphors and beliefs for specific roles.* Paper presented at the annual meeting of the American Education Research Association, San Francisco.

United States Department of Education (1993). *State of the art: Transforming ideas for teaching and learning science.* Washington: United States Government Printing Office.

Vygotsky, L. (1978). Mind in society: *The development of higher psychological processes* (M. Cole, V. John-Steiner, S. Scribner, & E. Souberman, Eds.). Cambridge, MA: Harvard University Press.

———. (1989). *Thought and language.* (A. Kozulin, Ed.), (rev. ed.). Cambridge, MA: The MIT Press.

Wiggins, G. (1989, May). A true test: Toward more authentic and equitable assessment. *Phi Delta Kappan*, 70(9), 703–713.

———. (1992, May). Creating tests worth taking. *Educational Leadership*, 49(8), 26–33.

Witt, H. (1995, May 14). Bad grades for new age education: Low scores may lead California back to old teaching methods. *The Chicago Tribune*, pp. Section 1-1, 1–10.

Wolcott, H.F. (1975). Criteria for and ethnographic approach to research in schools. *Human Organization*, 34(2), 111–127.

——. (1990a). On seeking-and rejecting-validity in qualitative research. In E.W. Eisner and A. Peshkin (Eds.), *Qualitative inquiry in education: The continuing debate* (pp. 121–152). New York: Teachers College Press.

——. (1990b). *Writing up qualitative research.* Thousand Oaks,CA: Sage.

——. (1994). *Transforming qualitative data: Description, analysis, and interpretation.* Thousand Oaks,CA: Sage.

Wood, T., Cobb, P., & Yackel, E. (1990, April 13). *"Reflections on learning and teaching mathematics in elementary school."* Paper presented at the seminar series Constructivism in Education at the University of Georgia.

Yager, R.E. (1991a). Science/technology/society as a major reform in science education: Its importance for teacher education. *Teacher Education,* 3(2), 91–100.

——. (1991b). The constructivist learning model: Toward real reform in science education. *The Science Teacher,* 58(6), 52–57.

Yager, R.E., & Lutz, M.V. (1994). Integrated science: The importance of "how" versus "what." *School Science and Mathematics,* 94(7), 338–346.

Yin, R.K. (1994). *Case study research: Design and methods* (2nd ed.). Thousand Oaks,CA: Sage.

Zemelman, S., Daniels, H. & Hyde, A. (1993). *Best Practice: New standards for teaching and learning in America's schools.* Portsmouth, NH: Heinemann.

APPENDIX A
Specific Data Collection and Analysis Methods

Data Collection. Data collection consisted of multiple data sources and included fieldnotes of daily classroom observations, teacher interviews, interviews with the school principal and secondary informants, informal conversations with other teachers, and collection of pertinent school and classroom documents. In addition, Donna reviewed and responded to the Science Teaching Efficacy Belief Instrument (STEBI) (Riggs & Enochs, 1990). Fieldnotes from classroom observations as well as all formal interviews (there were six in total) were transcribed onto a computer and printed out to attain a hard copy. Each classroom observation was audiotaped, and these were used to corroborate what the teacher said during the classroom lesson. In addition, and with time permitting, I audiotaped brief discussions with Donna about her prelesson goals and postlesson reflections. For BSCS unit two, she provided me with an audiotaped description of her plans and goals for the entire unit, as well as weekly written lessons plans for the first few weeks of the unit. There were numerous informal telephone and face-to-face conversations throughout the year, and I made every reasonable attempt to document and take notes from each of these. Unless otherwise noted, all quotes that appear in this text are from the classroom observations, and teacher interviews and were taken from audiotaped transcriptions. Every attempt was made to ensure accuracy of these transcriptions.

The first three interviews (#1, 10/20/93; #2, 11/3/93; & #3, 11/19/93) consisted of collecting background information about Donna's professional history, her beliefs about teaching, her beliefs about teaching science, her self-efficacy of teaching science, her views of the curriculum, and her expectations of the students. The fourth and fifth (#4, 12/2/94, & #5, 6/13/94) interviews focused on specific issues and questions centered on themes that emerged from the observational data. For example, interview four focused on her views of science fair and the problems and dilemmas she had with it. Interview five, which was a preactivity interview for "Oil Spill," also had questions that asked her to reflect back on her earlier statements about student expectations and to evaluate the students' progress based on her beliefs and accomplishments so far during the school year. Finally, interview six (#6, 7/7/94) was a culminating interview and asked her to briefly revisit her beliefs about teaching science, to discuss evidence of success, as well as to address specific issues regarding her classroom practice. In this session there where numerous occasions when Donna deviated a little from the initial question, using a story or an example to stress a certain point. All of this information, while useful in ascertaining who she is and what is important for her as a teacher and person, will not make its way into this document.

Data Analysis. Using the identified domains and patterns, I began to "funnel" (Spradley, 1980) my descriptive data into more specific categories. When I found detailed classroom sequences that supported the domains, I transcribed the classroom audiotapes to attain more exact information regarding Donna's discussions on the classroom structure. While student responses are desirable, location of the tape recorder and the volume level of student talk did not allow for accurate retrieval of student discourse. Thus, the bulk of the enactment component of the model is the descriptive data and the transcribed classroom tapes of teacher talk. This is consistent with the intent of the study.

Formal interviews were transcribed in their entirety. Within the initial domain of "Self," I coded and collected excerpts from four main categories. These were Donna's views of herself as a teacher, as a learner, as a science teacher, and as a philosopher of education. This domain was then relabeled "Teacher Beliefs," which allowed for the incorporation of more categories. These included Donna's beliefs about her students, beliefs about the curriculum, beliefs about science and its usefulness to her students, and beliefs about the world-at-large and the relationship between what happens in her classroom and what happens in life. Information about her professional background and the influences this had on where she is today were also coded. Overlap of the data categories was common. For example, her beliefs about teaching and teaching science were nearly one and the same. This information was then used to construct a profile of Donna's belief structure. The intent was to compare beliefs to actions in the enactment process and to gain an understanding of what happens in the classroom and why.

I used the principal interview as well as school documents to create the profile of the school and classroom context. Donna also provided me with demographic data specific to her students. Additional information came from other informal discussions and observations.

The profile on classroom curricula came from two primary sources-the written text and implementation guide of BSCS (1992a, 1992b) and Donna's account of the creation of the labs and science fair. Secondary sources, a principal writer for BSCS, and the woman in charge of the school implementation and in-service program, were consulted for more

Appendix A: Specific Data Collection and Analysis Methods

information about *Science for Life and Living*, the BSCS curriculum. The categories in the ensuing description were then taken from those in the curriculum itself. This information was then grouped into two specific categories that related to the instructional material and classroom structure.

Data analysis was ongoing. Eventually the process must cease and the descriptions must stand alone. What you will see is not an end to that process but my current best effort to show clearly and succinctly what happened in this classroom. Ideally, if another person were to organize this data, she would see similar patterns of how this teacher structured the learning in her classroom and how this structure was funneled through her belief system.

APPENDIX B
Donna's Responses to the Science Teacher Efficacy Belief Instrument

(STEBI). Riggs and Enoch, 1990)

Please indicate the degree to which you agree or disagree with each statement by circling the appropriate letters to the right of each statement.

SA=Strongly Agree
A=Agree
UN=Undecided
D=Disagree
SD=Strongly Disagree

maybe one? maybe not another!

1. When *a student* does better than usual in science, it is often because the teacher exerted a little extra effort.   SA   A   UN   D   SD

2. I am continually finding better ways to teach science.   SA   A   UN   D   SD

3. Even when I try very hard, I don't teach science as well as I do most subjects.   SA   A   UN   D   SD

4. When the science *grades* improve, it is most often due to their teacher having found a more effective teaching approach.   SA   A   UN   D   SD

5. I know the steps necessary to teach science concepts effectively.   SA   A   UN   D   SD

6. I am not very effective in monitoring science experiments.   SA   A   UN   D   SD

7. If students are underachieving in science, it is most likely due to ineffective science teaching.   SA   A   UN   D   SD

8. I generally teach science ineffectively.   SA   A   UN   D   SD

9. The *inadequacy of a student's* science background can be overcome by good teaching.   *This is offensive language, isn't it?*   SA   A   UN   D   SD

10. The low science achievement of some students cannot generally be *blamed* on their teachers.   *Who's assigning blame — and why?*   SA   A   UN   D   SD

11. When a low achieving child progresses in science, it is usually due to *extra attention* given by the teacher.   *What is meant by "extra attention?"*   SA   A   UN   D   SD

12. I understand science concepts well enough to be effective in teaching elementary science.   SA   A   UN   D   SD

The instrument is pretty transparent. I find the items that ask me to get on one side of the fence or the other to be, at best, confusing and poorly written. At their worst, I find them offensive. These items are written as if there can only be two bipolar explanations for why students achieve in science: us, or them. Is it unreasonable

[Handwritten annotation at top: "to suggest that student achievement in science is not about what the students do or about what the teacher does, but rather, isn't it about what we do together? I think that student achievement is more about the relationships that we build among students, teachers, content, and process. So, I can't answer these questions. And I won't... NYAAHHH!"]

13. Increased effort in science teaching produces little change in some students' science achievement.   SA   A   UN   D   SD

14. The teacher is generally responsible for the achievement of students in science.   SA   A   UN   D   SD

15. Students' achievement in science is directly related to their teacher's effectiveness in science teaching.   SA   A   UN   D   SD

16. If parents comment that their child is showing more interest in science at school, it is probably due to the performance of the child's teacher.   SA   A   UN   D   SD

17. I find it difficult to explain to students why science experiments work.   SA   A   UN   D   SD

[Handwritten: "These are really troubling!" and "why is this necessary? So we're supposed to be one right answer?"]

18. I am typically able to answer students' science experiments.   SA   A   UN   D   SD

[Handwritten: "What?!? Why should I have to? Aren't we supposed to be experimenting, now would they?"]

19. I wonder if I have the necessary skills to teach science.   SA   A   UN   D   SD

20. Effectiveness in science teaching has little influence on the achievement of students with low motivation.   SA   A   UN   D   SD

[Handwritten: "How can effectiveness in teaching be separated from motivation? Effectiveness in teaching is about motivation."]

21. Given a choice, I would not invite the principal to evaluate my science teaching.   SA   A   UN   D   SD

22. When a student has difficulty understanding a science concept, I am usually at a loss as to how to help the student understand it better.   SA   A   UN   D   SD

23. When teaching science, I usually welcome student questions.   SA   A   UN   D   SD

24. I don't know what to do to turn students on to science.   SA   A   UN   D   SD

25. Even teachers with good science teaching abilities cannot help some kids learn science.   SA   A   UN   D   SD

Science Teacher Efficacy Belief Instrument (STEBI) from Riggs, I.M., & Enochs, L.G. (1990). Toward the development of an elementary teacher's science teaching efficacy belief instrument. *Science Education*, 74(6), 625–637. Reprinted by permission of Wiley-Liss, Inc., a subsidiary of John Wiley & Sons, Inc.

Studies in the Postmodern Theory of Education

General Editors
Joe L. Kincheloe & Shirley R. Steinberg

Counterpoints publishes the most compelling and imaginative books being written in education today. Grounded on the theoretical advances in criticism, feminism, and postmodernism in the last two decades of the twentieth century, Counterpoints engages the meaning of these innovations in various forms of educational expression. Committed to the proposition that theoretical literature should be accessible to a variety of audiences, the series insists that its authors avoid esoteric and jargonistic languages that transform educational scholarship into an elite discourse for the initiated. Scholarly work matters only to the degree it affects consciousness and practice at multiple sites. Counterpoints' editorial policy is based on these principles and the ability of scholars to break new ground, to open new conversations, to go where educators have never gone before.

For additional information about this series or for the submission of manuscripts, please contact:
>Joe L. Kincheloe & Shirley R. Steinberg
>c/o Peter Lang Publishing, Inc.
>275 Seventh Avenue, 28th floor
>New York, New York 10001

To order other books in this series, please contact our Customer Service Department:
>(800) 770-LANG (within the U.S.)
>(212) 647-7706 (outside the U.S.)
>(212) 647-7707 FAX

Or browse online by series:
>www.peterlangusa.com